MAKING USE
BIOLOGY

for GCSE

MAKING USE OF
BIOLOGY

for GCSE

MAKING USE OF BIOLOGY

for GCSE

Pauline Alderson
and
Martin Rowland

MACMILLAN

First published 1989 by
MACMILLAN EDUCATION LTD
Houndmills, Basingstoke, Hampshire RG21 2XS
and London
Companies and representatives
throughout the world

ISBN 0–333–45992–X

Printed in Hong Kong

Reprinted 1991, 1992

M

MACMILLAN

CONTENTS

ACKNOWLEDGEMENTS

The authors and publisher wish to acknowledge the following photograph sources:

Bernard Alfieri p. 143 (top right);
Heather Angel pp. 96, 131 (left), 144, 145, 146;
Arthur Bishop p. 76;
Camera Press pp. 74, 244;
J. Allan Cash p. 209;
Courtaulds p. 23;
Gene Cox pp. 99 (top and bottom), 166, 168 (top), 322;
Family Planning Association p. 270;
Food and Wine from France p. 72;
Glaxo Holding plc p. 89;
Richard and Sally Greenhill pp. 220, 328;
Health Education Council p. 286;
John Innes Institute pp. 132 (photo C. Hussey), 309;
Popperfoto p. 4;
RHM Research p. 90;
Science Photo Library pp. 16 (photo Dr Jeremy Burgess), 18 (photo Eric Grave), 20, 98, 177 (photo Dr Jeremy Burgess), 143 (top left), 179 (photo Dr Howells), 234, 307, 318;
Topham Picture Library p. 156;
Turberville Smith and Son p. 212;
R. Upptograffe p. 83;
Watney Brewery p. 67;
C. James Webb p. 168 (bottom).

The publishers have made every effort to trace all the copyright holders, but if they have inadvertently overlooked any, they will be pleased to make the necessary arrangements at the first opportunity.

254, 307, 318;

EXPERIMENTS

Part 1

Part 2

x

INTRODUCTION FOR TEACHERS

This book concentrates on "topics related to the personal, social, economic and technological applications of biology." In order to satisfy the *National Criteria*: *Biology*, a minimum of 15 per cent of all GCSE Biology Syllabuses is devoted to these topics, on which the traditional textbooks give little or nothing. This book has the emphasis on biotechnology required by the *National Criteria*.

It covers the London and East Anglian Group's Syllabus for the GCSE Biology (Series 17) examination, including the practical work. Though originally described as the 'Mature' syllabus, this is available at any age and many schools have adopted it as their mainstream course because of its intrinsic interest. The reason there is no separate Human Biology (Series 17) examination is that it could differ only slightly from the one covered in this book. Not only does the LEAG's Biology (Series 17) Syllabus include much human biology but a Human Biology (Series 17) Syllabus, in order to satisfy the *National Criteria*, would have to include much basic biology.

At least a week can be allowed for each Unit if the course is taught in a year, two weeks if it is taught in two years. Unit 6 on 'Fermentation' and Unit 25 on 'Genetics, Cell Division and Genetic Engineering' may need twice as long as most Units.

Part 1 deals with 'Economic and Environmental Biology', Part 2 with 'Human and Social Biology'. Either Part can be taught first. Students who have done no Biology before might find it easier to begin with Part 2, where the early Units cover some basic subject-matter and have experiments that are easier than most. Many students coming to this Syllabus will have a basic biology background even if it is in 'Science': the early Units in Part 1, because their subject-matter is unfamiliar and up to date, are more likely to capture their attention and interest.

No attempt has been made to put an equal amount of practical work at the end of each Unit. The Experiments come at the ends of the Units to which they are relevant. Whereas Unit 8 has four Experiments, a number of Units have none. The Experiments should be spread throughout the course. They may be done in any order except where a 'Note to teachers' says not.

You should read not only Appendix D on 'Practical Work' but also the guidance on Individual Studies given to students in Appendix B, as well, of course, as the Syllabus. In Appendix B we advise students to choose for their Individual Studies experiments that test a hypothesis. The LEAG (Series 17) Syllabus says that an experiment should include a statement of the "aim, problem or hypothesis" and that "Candidates should have the opportunity to formulate a hypothesis or state a problem to be examined" but adds "and to design an experiment or investigation to test it." Hypotheses lend themselves to testing whereas problems do not. Moreover the criteria for assessment for high standard and mid standard both assume a hypothesis; only the criteria for low standard make no mention of a hypothesis.

At the end of each Unit there are at least three questions of varying difficulty. Because students cannot have too much practice in answering suitable questions, it is desirable to supplement those in this book with past examination papers or specialist books of questions.

This book uses the Institute of Biology's recommendations on biological nomenclature, units and symbols for secondary education. It uses billion to mean a thousand million (1 000 000 000). For the sake of clarity it hyphenates compound adjectives including chemical compounds when they are used as compound adjectives.

Nothing in this book requires students or teachers to kill animals. Animals brought into the classroom for study should be returned unharmed to their habitats.

INTRODUCTION FOR STUDENTS

This book tells you how yoghurt, beer and wine are made. It tells you about biological washing powders, contraception, test-tube babies, heart attacks, AIDS, drugs and new forms of food. It describes how criminals can be identified from a piece of skin, a hair root, a drop of blood or (in rape cases) a drop of semen; how smoking and drinking affect health; how we can find out if our children will be born with handicaps or diseases; how tiny organisms can be used to make things for us; how ladybirds can be used to eat the pests on crops, with the result that the pests need not be killed by chemicals that end up in our food.

It is unlike most biology textbooks because it is largely about humans and real life. Remember that what goes on in real life changes all the time. Cut out any stories you see in newspapers about new food or new methods of food production, new drugs, new methods of contraception, new dangers from pollution or anything else that involves something biological.

All the Units in this book have at least three questions at the end: the pages on which there are questions are flagged, which makes it easy to find them. The questions are all of the kind you will get in your GCSE examination. At the end of each Unit, the easy questions usually come first and the difficult ones last.

There is practical work at the end of many of the Units. The practical work includes all the 23 experiments you need to know for the LEAG Biology (Series 17) examination. Four experiments that you design yourself have 'Clues to help you design your experiment'. These clues are in tinted red boxes, as are passages that affect safety. The experiments you design yourself are good practice for the two Individual Studies you need to do during the LEAG course. These are two experiments which you choose, design and do yourself and write reports about. They are assessed by your teacher as part of your GCSE examination. Appendix B helps you with your Individual Studies.

Appendix A tells you the bits of mathematics, physics and chemistry you need to know. Appendix C tells you how to draw and compare biological specimens, how to answer examination questions and how to revise.

PART 1
Economic and Environmental Biology

VARIETY OF LIVING THINGS

Next time you are walking out of doors, look at the plants and animals around you. Even in the towns you will see trees, shrubs, grass, moss, a few flowers, birds, dogs, cats, humans and occasionally horses. If you search, you may find butterflies, moths, bees, flies, earwigs, spiders, earthworms and snails. We call all these living things **organisms**. All around you are millions more organisms too small for you to see. We make use of many organisms throughout our daily lives: we could not live without them.

(a) Mammals

The animals we know best are those in the group we belong to ourselves, the **mammals**. Other mammals are elephants, lions, dogs, cats, squirrels and mice. Mammals are different from all other living organisms in a number of ways. They have **hair**, which on most of them we call **fur**. They usually have **outer ears**, which means part of their ears stick out from the rest of their bodies. They have several **different** kinds of **teeth** in their mouths, whereas other animals with teeth, such as fish, have all the same kind. Female mammals have glands on their chests or abdomens ending in **nipples** from which their young drink milk. (Male mammals have nipples too, but in most species they do not release milk.) The word meaning *on the breast* is *mammary*. These milk-producing glands are therefore called **mammary glands**. Mammals take their name from these mammary glands because the fact that they suckle their young is the most unusual thing about them.

We know so much about mammals because we need them in our everyday lives. We need them to give us food and clothing: cattle and goats give us not only meat and milk but leather; sheep give us not only meat, milk and leather but wool. People ride horses, mules, camels and oxen or use them for drawing carts and ploughs. Horses also run races for our sport. Figure 1.1 shows a dog rounding up sheep; we also keep both dogs and cats for company. Because mammals are so important to us, we long ago learned to look after them and breed them.

(b) Birds

Birds are different from all other living organisms because they have **feathers** and a **beak**. Chickens, ducks and geese give us meat and eggs. Pigeons carry messages and take part in races for us. Canaries and budgerigars keep us company and

Figure 1.1   Sheepdog and sheep

please us with their singing. Because birds, like mammals, are important to us, we have learned to look after and breed them too.

(c) Fish

Fish have **streamlined bodies** with powerful **muscular tails** that help them to swim. They have two pairs of **fins** on the sides of their bodies as well as fins along the tops and bottoms of their bodies. They have a sense organ along each side of their body by which they feel movements in the water; because the word for *on the side* is *lateral*, this sense organ is called a **lateral line**. Fish breathe by taking water into their mouths and passing it over their **gills**, which take oxygen from it as our lungs take oxygen from air. People have been catching and eating fish for millions of years, and they have learned how to look after them and breed them. There are trout farms all over the UK.

(d) Vertebrates

Mammals, birds and fish look different and live differently. But they all have **vertebral columns** (backbones) and therefore all belong to the same large group of **vertebrates**. Other examples of vertebrates are **reptiles**, such as snakes and lizards, and **amphibians**, such as frogs, which can live both in water and on land.

(e) Other animals

Animals without vertebral columns are much smaller than vertebrates. Many have been caught, not bred, by people who have wanted to eat them or make other use of them. Now crayfish, mussels and snails are being farmed for food, while for centuries silkworm moths have been reared on mulberry bushes for their silk.

1.2 Classification

Because there are so many different organisms, we need to sort them out, to put them in groups, that is to **classify** them. As long ago as the fourth century BC Aristotle made a classification of living organisms. Two thousand years later a

4

Swedish botanist, Carl Linnaeus (1707–78), started a system of naming organisms which we still use today. This system gives every organism two names. We ourselves are called *Homo sapiens*. *Homo* is the Latin for *man*, that is for a *human*, and *sapiens* is the Latin for *wise*. The first name, which always begins with a capital letter, is for the larger of the two groups and is called the **genus**. The second name, which always begins with a small letter, is for the smaller of the two groups and is called the **species.**

A species includes organisms which biologists have decided are similar enough to be classified in the same group. Organisms that reproduce sexually are classified in the same species if they can interbreed (in natural conditions) and their offspring are fertile. Horses and donkeys belong to different species: though they can be interbred, their offspring, mules and hinnies, are infertile. The white, brown, black and yellow races of humans are all one species, *Homo sapiens*, because we can all interbreed with one another. We are the only species of the genus *Homo* still living today. We are called *Homo sapiens* because we are more intelligent than were the other species of *Homo*.

The name for dog is *Canis familiaris*. On the continent of Europe there are other species of the genus *Canis* living wild: the wolf, *Canis lupus*, and the jackal, *Canis aureus*. An Alsatian dog, a wolf and a jackal look much alike and have been known to interbreed in captivity, but they do not interbreed in the wild. This is why they are three different species. The dogs we know are very different from the wild dogs they have descended from. This is because people decided to make them different. Dogs are the result of **selective breeding**.

Dogs belong to the canine family (a group including foxes and wolves); canines belong to a larger group, the **carnivores**, that belong to the mammals; mammals belong to the vertebrates, and vertebrates belong to the animals. The animals are one of the five kingdoms into which all living organisms are now divided:

Animalia (animals)
Plantae (plants)
Protoctista (protoctists)
Fungi (fungi)
Prokaryotae (prokaryotes)

We have already looked at the animals. You will see that we also need the organisms in all the other four kingdoms.

1.3 Plants

Plants are organisms that are green in colour and make their own food from air, water and simple inorganic ions. They usually have roots, stems and leaves. Humans have always used plants as food. At first they ate only plants growing wild. About twelve thousand years ago the farming revolution began. People began to grow crops. At first they collected seeds of wild wheat, oats and barley and sowed them in soil they had prepared so that they would grow well. Then they took seeds from the crops they had grown and sowed them too in prepared

5

soil. People were no longer wandering about in search of food but were staying in one place and growing their food. They first used and then bred oxen to pull carts and ploughs. They built permanent houses. They also built walls to keep both wild animals and their own animals away from their crops. Villages were formed.

In the UK we now grow all sorts of crops. We still grow wheat, oats and barley, but the varieties we use give a much bigger yield than those of twelve thousand years ago. This is the result of selective breeding of wheat, oats and barley (see Section 11.4). We grow other cereal grains too, such as rye and maize. In warmer climates people grow rice and millet. It is the seeds of all these crops which people eat.

We eat other parts of plants: the leaves of cabbages, lettuces, corianders and onions; the roots of carrots, parsnips and yams; the fruits of apples, beans and chillies; the stems of potatoes and sweet potatoes; the flowers of cauliflowers and broccoli. Sugar, tea and coffee are also products of plants. Not all of these are grown as crops in the UK; some are grown in warmer climates. In Figures 9.6 and 26.4 you can see some of the many plants that we eat.

Some plant crops are grown for uses other than food. Trees are grown for timber for buildings, furniture and paper. The cotton plant is grown for cotton to make into cloth. Figure 1.2 shows a number of products of living organisms.

There are bread, cheese, yoghurt, pepper, salad, pickled onions, beer and wine to eat and drink. Do you know what living organisms are involved in making these products? What products of living organisms can you see in the photograph apart from the food and drink? What do these products come from? Though you cannot tell it from the photograph, the curtains are the product of a living organism. What might that living organism be?

1.4 Protoctists

Some protoctists are animal-like; others are plant-like; a few can change from

6   Figure 1.2   Products of living organisms

being animal-like to being plant-like. Most are so small they can be seen only under a microscope. The larger ones, such as **seaweeds**, do not form roots, stems or leaves. In some parts of the UK people have eaten protoctists for centuries. In south Wales they mix the red seaweed *Porphyra* with oatmeal to make a cake called laverbread, which is fried. The green seaweed *Ulva*, known as green laver or sea lettuce, has also been eaten for centuries.

In Japan nearly 100 000 people are employed cultivating and processing protoctists from the sea. In Balmuchy in Scotland the Sea Vegetable Company (*sea vegetable* sounds better than *seaweed*) is marketing seaweeds which can be eaten raw in salads, made into sauces or added to fish soups. Some protoctists can be used for making paper.

1.5 Fungi

Fungi are plant-like because they do not move about, but they are never green like plants. Many are small and cannot be seen without a microscope: **yeast**, which people have used for more than 6000 years to make wine and beer, is a small fungus (singular of *fungi*). Although people used it, they never saw what it was until microscopes were invented 300 years ago; even then they did not realise it was a living organism.

Mushrooms, which we eat raw or cooked, are fungi that are large enough for us to see without a microscope. So are the fungi that form white cotton-wool-like threads over bread and fruit if they go bad. Most fungi have bodies that are made up of white threads, but yeast does not (see Figure 2.2).

Yeast and mushrooms are not the only fungi in our food. Many cheeses are made by the action of fungi on milk. Blue-vein cheeses have an internal growth of the fungus *Penicillium*. Farm animals have been successfully fed on **mycoprotein**, a food made from fungi, for many years. Humans are eating mycoprotein too now that manufacturers have made it look and taste like meat.

In the 1940s we discovered another use for fungi: the production of antibiotics, which doctors use to kill bacteria that have made people ill. The first antibiotic was **penicillin**, made, like blue-vein cheeses, from the fungus *Penicillium*. Now there are many different antibiotics made from several fungi.

1.6 Prokaryotes

Prokaryotes are too small for us to see without a microscope. They are smaller than protoctists and fungi. Even under a good light microscope they often look no bigger than specks of dust. It is only since the electron microscope was invented in the 1920s that we have been able to see what prokaryotes look like. All **bacteria** (singular *bacterium*) are prokaryotes and so are some very small blue-green plant-like microorganisms.

Four centuries ago Spanish explorers found that the Mexican Aztecs were farming on the surfaces of lakes a green 'ooze' which they made into cakes with a cheese-like flavour. This ooze is a prokaryote called *Spirulina*, which the people of Chad in Africa have also grown for centuries. In Europe in recent years factories have been built to produce high-quality protein food from *Spirulina*. (Protein is the part of food that all organisms need in order to grow.) The food that they are making is called **single-cell protein** or **bug-grub**.

Bacteria have long been used to make cheese and yoghurt for us. Since the 1980s bacteria have also been used to make medical drugs by what is called **genetic engineering** (see Section 25.4).

1.7 How to identify an organism

No one knows the names of all the organisms in the world. Even the experts on insects do not know the names of all the insects. There are books to help people identify organisms. It may be possible to identify an organism from a photograph, a drawing or a description. Another way is to use a **key**.

(a) Using a key

Suppose you want to name the organism shown in Figure 1.3. You can use the key in Figure 1.4. To use a key you start at description 1 and read the alternatives: the simplest keys have just two alternatives in each description. You look at the organism you want to identify and see which of the alternatives fits it.

When you have decided which alternative fits, you look along the line to the right where you will find either a number which tells you which description to read next or the name of your organism.

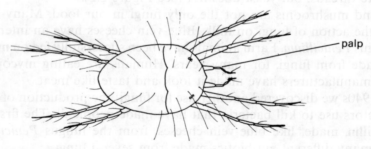

Figure 1.3   Unknown organism

1	Legs present	2
	No legs	5
2	3 pairs of legs	fly
	More than 3 pairs of legs	3
3	4 pairs of legs	4
	7 pairs of legs	woodlouse
4	Bristles on body; no waist	mite
	No bristles on body; waist present	spider
5	Shell present; 2 pairs of tentacles	snail
	No shell; no tentacles	6
6	Body surface segmented	earthworm
	Body surface not segmented	eelworm

Figure 1.4   Key

The organism in Figure 1.3 has legs: after reading description 1 you find the number 2 at the end of the line with your chosen alternative. You go to description 2. Because your organism has more than three pairs of legs, you go to description 3. At first you might think your organism has five pairs of legs. In fact it has only four: the front pair of what may look like legs are palps with which it touches things. Here you have a good reason why it is better to see the organism itself than a picture of it: it would not be walking on its palps. Even so you are unlikely to make a mistake. The other alternative in description 3 is '7 pairs of legs' and your organism certainly does not have those. In all good keys the alternatives are quite different from one another. Now you go to description 4 and the end of the trail. The organism is a **mite**. It is often called a red spider; it can cause a lot of damage to plant crops by sucking their sap.

Look at the animals shown in Figure 1.5. They are animals you can find in most gardens. See if you can identify them using the key in Figure 1.4. They have been given letters, not names, in the diagram: try to track down the name for each letter. Although you will recognise most of them, the answers are given at the end of this Unit.

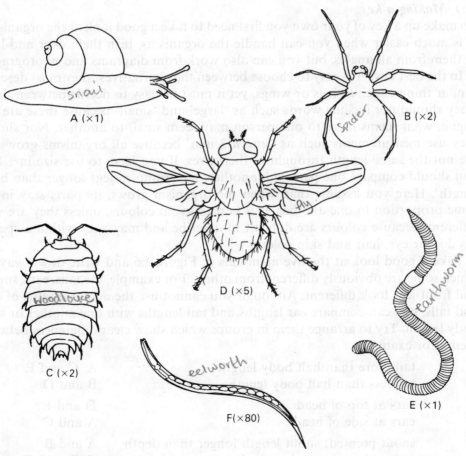

Figure 1.5 Garden animals

9

(b) Magnification

After the letter for each animal in Figure 1.5 there is a bracket with a multiplication sign and a number. This is a way of telling you the approximate real-life size of what has been drawn. It gives the scale or the **magnification** of the drawing. In the drawing animal D is five times longer than it is in real-life: it is therefore a small animal, less than 1 cm long. Animal F in real-life is even smaller, less than 1 mm long.

Sometimes we need to draw animals smaller than they are in real-life. All the animals in Figure 1.6 are drawn smaller than they are in real-life. The phrase 'not drawn to scale' means that they are not all drawn to the same magnification. Because they are smaller than they are in real-life, the scale is shown as a decimal fraction. For example, the rabbit is about one-tenth of its real-life size: this would be written as ($\times 0.1$). The mole is about a third of its real-life size: this would be written as ($\times 0.33$). Try working out the magnifications of: the common shrew, which has a head-and-body length (excluding the tail) of about 8.25 cm; the bank vole, which has a head-and-body length of about 10.5 cm; the grey squirrel, which has a head-and-body length of about 22 cm. The answers are given at the end of this Unit.

(c) Making a key

To make up a key of your own you first need to take a good look at the organisms. It is much easier when you can handle the organisms, turn them over and look at them from all angles, but you can also work from diagrams and photographs.

In the best keys it is easy to choose between the alternatives: they must describe similar things, such as legs or wings, yet it must be easy to decide between them. They should not include words such as 'large' and 'small' because these are too vague: what seems large to one person may seem small to another. Nor should they use measurements such as 'length 3 mm' because all organisms grow and are not the same length throughout their lives. If you have to use size in a key, you should compare one part with another: for example, 'legs longer than body length'. Here you assume that, when the organism grows, its parts stay in the same proportion to one another. It is best to avoid colours, unless they are very different, because colours are difficult to describe and may vary within a species (as do our eye, hair and skin colours).

Take a good look at the five mammals in Figure 1.6 and write down ways in which some are obviously different from others. For example, the tails, ears, snouts and front legs look different. Although you cannot use the actual lengths of ears and tails, you can compare ear lengths and tail lengths with one another or with body length. Try to arrange them in groups which show clear differences between them. For example:

tail more than half body length	A, C and E
tail less than half body length	B and D
ears at top of head	D and E
ears at side of head	A and C
snout pointed; snout length longer than depth	A and B
snout blunt; snout length shorter than depth	C, D and E

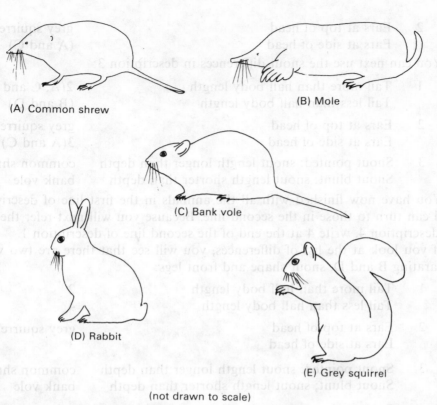

(A) Common shrew

(B) Mole

(C) Bank vole

(D) Rabbit

(E) Grey squirrel

(not drawn to scale)

Figure 1.6   Five British mammals

front legs as broad as or broader than long	only B ✓
front legs longer than broad	A, C, D and E

There are several different ways in which we can arrange these differences to make a key. You may like to try to make your own before reading on. You may start with

1	Tail more than half body length	2(A, C and E)
	Tail less than half body length	(B and D)

While you are making your key, you may find it helps to put 'A, C and E' in brackets at the end of the first line and 'B and D' in brackets at the end of the second line. Remember to leave them out when you write your key out neatly.

Because you will next refer the reader to description 2, write 2 at the end of the first line. Do not yet number the second line.

To separate A, C and E into two groups you cannot use the difference in the front legs because all three have legs that are longer than they are broad,. You can, however, use the ear and snout differences. Suppose you first use the ear differences in description 2:

1	Tail more than half body length	2(A, C and E)
	Tail less than half body	(B and D)

11

| 2 | Ears at top of head | grey squirrel |
| | Ears at side of head | (A and C) |

You can next use the snout differences in description 3:

1	Tail more than half body length	2(A, C and E)
	Tail less than half body length	(B and D)
2	Ears at top of head	grey squirrel
	Ears at side of head	3(A and C)
3	Snout pointed; snout length longer than depth	common shrew
	Snout blunt; snout length shorter than depth	bank vole

You have now finished with all the animals in the first line of description 1 and can turn to those in the second line. Because you will next refer the reader to description 4, write 4 at the end of the second line of description 1.

If you look at the list of differences, you will see that there are two ways of separating B and D, snout shape and front legs:

1	Tail more than half body length	2
	Tail less than half body length	4
2	Ears at top of head	grey squirrel
	Ears at side of head	3
3	Snout pointed: snout length longer than depth	common shrew
	Snout blunt; snout length shorter than depth	bank vole
4	Snout pointed; snout length longer than depth; front legs as broad as or broader than long	mole
	Snout blunt; snout length shorter than depth; front legs longer than broad	rabbit

You have now completed the key. Paragraph 4 is a good one because it uses three differences instead of only one. Where you can use two, or even more, differences, it makes for a better key.

Questions

Q 1.1 Look at the three different mammals in Figure 1.1. What can you see that tells you they are mammals?

Q 1.2 List the five kingdoms of living organisms. Describe a way in which humans have used **one** organism from each kingdom to improve their lives.

Q 1.3 Describe **three** of the *early* ways in which humans used other organisms to improve their lives.

Practical work

Experiment 1.1   To use an identification key
Note to teachers

12   You may substitute the photographs in Section 12.2 for the animals listed but

students will not be able to find all the features unless they are able to handle the animals and turn them over.

Specimens can be obtained from biological suppliers but most can be found in gardens and college or school grounds. Aphids can usually be found on rose bushes (greenfly) and bean plants (blackfly), where you may also find ladybirds feeding on them. Centipedes and millipedes can be dug from fertile soil. You can encourage earthworms to leave their burrows by watering the soil from shoulder height. If you place a half grapefruit or orange, flesh downwards, in a garden, grassland or woodland overnight, you should find slugs inside it in the morning.

You might obtain dead honey bees from a local bee keeper. (Live honey bees have a painful sting.)

Using living organisms may be distasteful to some students and will cause problems when they move. On the other hand, the use of organisms killed for the purpose of experiment or study may be held to be incompatible with Aim 2.1.4 ('To promote respect for all forms of life') of *The National Criteria: Biology*.

Materials needed by each student
1 pair of forceps
1 hand lens
1 aphid
1 centipede
1 earthworm
1 honey bee
1 ladybird
1 millipede
1 terrestrial slug

Instructions to students
1. Use the key to name each of the animals you have been given. You may need to move the animals to look at them from different angles or to look at them through the lens. 'Legs 0' is another way of writing 'No legs'.

Key

1 Legs 0; rigid surface 0	2
Legs present; rigid surface present	3
2 Head with tentacles present; body segments 0	slug
Head with tentacles 0; body segments present	earthworm
3 Legs 6 or less; wings may be present	4
Legs more ·than 6; wings 0	6
4 Wings 0 or all wings fragile and transparent	5
Front wings hard, leathery and not transparent	ladybird
5 Cornicles (little tubes) on each side of rear of body; body and legs smooth	aphid
Cornicles 0; body and legs 'hairy'	honey bee

13

6 One pair of legs per body segment   centipede
   Two pairs of legs per body segment   millipede

2. Some of the animals are very brightly coloured. Explain why their colour was
   not used in the identification key.

3. Make a list of some other features of the animals which have not been used
   in the key. Use these, and some of the features listed above if you need to, to
   make another key to distinguish between the same animals.

Answers to questions

The animals in Figure 1.5 are: A snail; B spider; C woodlouse: D fly; E earthworm;
F eelworm.
The scales of the mammals in Figure 1.6 are: common shrew (× 0.4); bank vole
(× 0.33); grey squirrel (× 0.15).

14

MICROORGANISMS

2

2

2.1 Introduction

Any organism so small that it can be seen only with a microscope is a **microorganism**. Microorganisms are found in three of the kingdoms: protoctists, fungi and prokaryotes. All prokaryotes are microorganisms. When light microscopes were invented three hundred years ago, people saw microorganisms for the first time and saw them only as small specks. Even when light microscopes improved, it was still difficult to see much inside a microorganism. When electron microscopes were invented in the 1920s, we learnt a great deal more about microorganisms.

2.2 Yeast (a fungus)

Yeast (genus *Saccharomyces*) is an organism, a **fungus**, less than 0.01 mm (10 μm) long, which humans have used for centuries to make alcoholic drinks and bread. Yeast is found growing naturally on the sticky sugary surfaces of ripe fruits and on most leaves.

(a) Structure

Yeast consists of a small ovoid unit called a **cell**. *Ovoid* means oval and three-dimensional like a plum. A cell is the smallest unit of living matter. Large organisms are made up of millions of cells but yeast consists of only one cell. Figure 2.1 is a photograph of dozens of yeast cells, taken with a special **scanning** electron microscope which shows in detail what is on the surface but not what lies inside.

Each cell of yeast includes **cytoplasm**, part of the living material, surrounded by a non-living wall, the **cell wall**. It is in the cytoplasm that most of the cell's living processes are carried out. **Food stores** lie in the cytoplasm. The surface of the cytoplasm forms a very thin membrane, the **cell-surface membrane**, inside the cell wall. Inside the cytoplasm is a compact body, the **nucleus**, which is surrounded by its own very thin envelope, the **nuclear envelope**. The nucleus, which contains **DNA** inside a number of **chromosomes** (see Section 25.3), controls the activities that take place in the cell. At the centre of the cell is a fluid-filled space, the vacuole. Food stores in the cytoplasm are usually in the form of droplets of lipid (oil) and granules of glycogen.

In Figure 2.1 you can see that most of the cells have outgrowths, called **buds**, by which they grow and eventually multiply, the buds breaking off to form new yeast cells. Figure 2.2 shows the inside structure of a yeast cell and a bud: the nucleus is dividing.

15

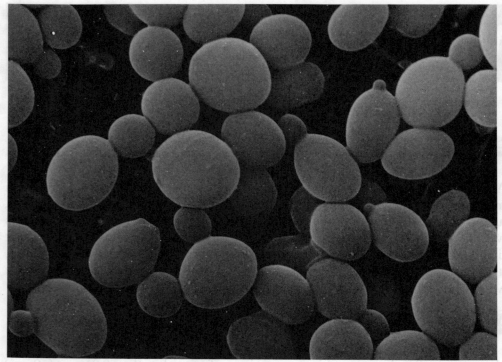

Figure 2.1 Yeast budding (× 2000)

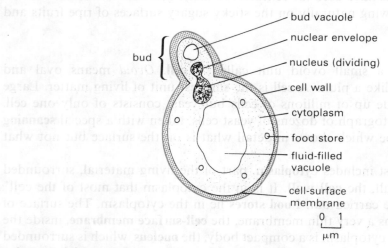

bud {

bud vacuole

nuclear envelope

nucleus (dividing)

cell wall

cytoplasm

food store

fluid-filled vacuole

cell-surface membrane

0 1
μm

Figure 2.2 Yeast contents

(b) Growth and reproduction

Like any other living organism, yeast needs a supply of food. Yeasts grow well on sugar but not on starch. They also need simple inorganic ions. Given suitable food and surroundings, yeast will make more of itself (more cytoplasm, nucleus and cell wall) in the form of a bud which grows from its surface. If conditions for growth are very good, more than one bud can form at a time, or a bud can itself form another bud before it separates. A bud breaks off from the parent yeast cell when the cell walls are complete between them. One of the cells in Figure 2.1 shows a clear **bud scar** where a bud has recently broken off. Can you see it? When the cells have separated, yeast has multiplied or **reproduced**.

2.3 Bacteria (prokaryotes)

The prokaryotes are all microorganisms. **Bacteria**, some of which cause us diseases, are prokaryotes. Most bacteria are less than 0.01 mm (10 μm) long. Figure 2.3 shows a typical bacterial cell. Although it has a cell wall, cytoplasm and food stores, it does not have a nucleus or vacuole. It has one long circular chromosome containing one long circular strand of DNA (see Figure 25.4). Some bacteria have an extra smaller circular piece of DNA called a **plasmid**: both the chromosome and the plasmid lie in the cytoplasm. Many bacteria have long threadlike strands, **flagella** (singular *flagellum*), which they use for swimming.

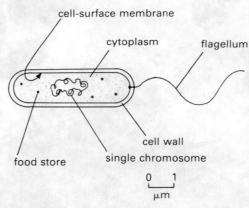

Figure 2.3   A typical bacterium

2.4 Rhizopus (a fungus)

Because yeast is so small, and because it is a single cell, it is an unusual fungus. *Rhizopus* looks very different from yeast. Like most fungi, *Rhizopus* consists of long threadlike strands, called **hyphae**. A hypha (singular of *hyphae*) has an outer wall, a large central fluid-filled vacuole and, lining the wall, a thin layer of cytoplasm which contains nuclei (plural of *nucleus*) and food stores. The fluid-filled vacuole stops the hypha collapsing: there are often no cross-walls to support it. The structure of a hypha is shown in a thin cut-through section in Figure 2.4.

The tangled mass of hyphae is called a **mycelium** or in everyday language a **mould**. A mycelium of *Rhizopus* is shown in Figure 2.5. You can see hyphae supporting black blobs above the mass of the mycelium, which is so tangled and dense that it is impossible to see separate hyphae. The black blobs are the parts of the fungus concerned with reproduction.

17

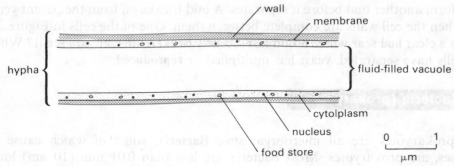

Figure 2.4   Hypha of a fungus

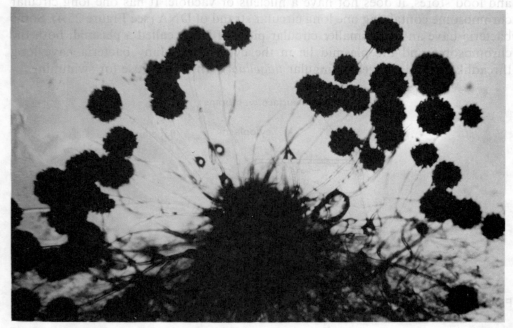

Figure 2.5   *Rhizopus nigricans* (× 60)

When bread and fruit go mouldy, it is usually because *Rhizopus* is growing on them. The moulds are mycelia (plural of *mycelium*). You can see tiny black blobs, the reproductive bodies, on the mycelia even without a magnifying glass. The mycelia of most fungi reproduce by forming millions of small cells, called **spores**, which are present in vast numbers in the air that surrounds us. Whenever a spore lands on suitable food, it grows to form the threadlike mycelium of the fungus.

(a) Growth

Figure 2.6 shows a fungal mycelium that has grown in the 24 hours after a spore landed on suitable food. You can see that the hyphae grow outwards in all directions after the spore lands on the food. A new wall and new cytoplasm and nuclei form behind each tip as each hypha grows outwards.

(b) Reproduction

18   The hyphae of different fungi are very much alike. Even experts have difficulty

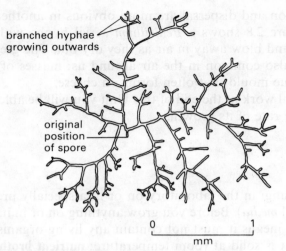

branched hyphae
growing outwards

original
position
of spore

0   1
mm

Figure 2.6   Mycelium after 24 hours' growth

identifying fungi just by looking at their hyphae. It is easier to identify fungi from the different ways in which they reproduce. *Rhizopus* reproduces a few days after starting to grow on food. Instead of growing outwards on the surface of their food, some *Rhizopus* hyphae grow into the air. Each aerial hypha swells at its tip, and rounded cells, the spores, form inside the tip. The tip is what you see as a black blob with your naked eye. It is covered in small black crystals, which, as in Figure 2.5, stop you seeing the spores inside. Figure 2.7 shows you how the spores form inside the swollen tip and how, as the swollen tip dries out, the wall breaks up, the tip collapses and the dry spores are **dispersed** (scattered) by being blown away into the air. When a spore lands on suitable food, it will form a new mycelium.

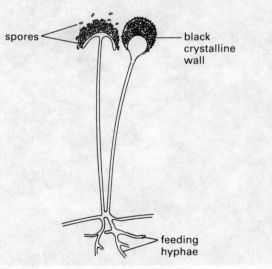

spores

black
crystalline
wall

feeding
hyphae

Figure 2.7   Spore formation and dispersal in *Rhizopus*

Spore formation and dispersal are more obvious in another common fungus, *Penicillium*. Figure 2.8 shows a *Penicillium* aerial hypha with chains of spores which separate and blow away in air as they dry out. The greeny-blue spores of *Penicillium* are also common in the air around us: masses of them can be seen as the greeny-blue mould that often forms on cheese.

In the practical work at the end of this Unit you will be able to grow *Rhizopus* and measure the rate of its growth.

You will grow fungi in the laboratory on or in a specially prepared food called a **medium** (plural *media*). Before you grow anything on or in it, the medium must be **sterile**, which means it must not contain any living organisms. **Nutrient agar** is a medium that is solid at room temperature; **nutrient broth** is a medium that is liquid at room temperature. When you add a microorganism to a medium, you **inoculate** the medium; the small quantity of the microorganism that you add to

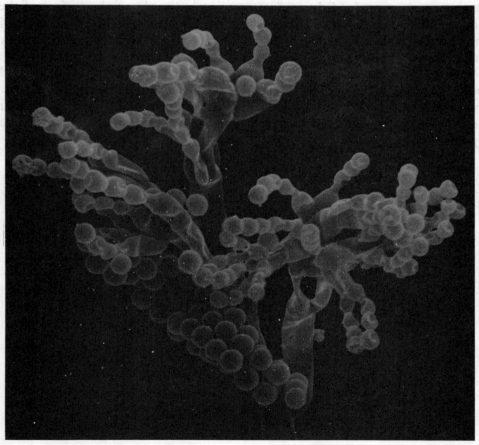

20   Figure 2.8   *Pencillium* (× 1300)

the medium is called an **inoculum**. For growth to take place, the inoculated medium must be kept at a suitable temperature for a suitable period of time: this is **incubation**. Incubation is usually carried out in a thermostatically controlled cabinet or **incubator**. Microorganisms grown from a single cell or mycelium are a **colony**. The medium and the microorganisms growing on or in it are a **culture**. A **pure culture** contains only one species of microorganism; a **mixed culture** contains more than one species. You will be using these terms throughout the early Units and will soon get used to them.

2.6 Safety

The practical work in the early Units involves growing microorganisms in the laboratory while taking strict safety precautions. Unlike many microorganisms, the ones you will deal with, yeast and *Rhizopus*, are harmless. This makes it possible for you to learn the safety precautions without danger.

2.7 Aseptic techniques

The liquid and solid media which you use to grow yeast and *Rhizopus* are suitable for the growth of many other microorganisms. Special methods, called **aseptic techniques**, must be used when transferring media and microorganisms to make sure that cultures do not become **contaminated** (spoilt by unwanted organisms).

(a) Sources of contamination
Microorganisms exist both as active cells and as **dormant** (inactive) cells: **spores** are dormant cells produced by many bacteria and fungi. Both as spores and active cells, microorganisms are blown into a laboratory through doors and windows: they grow only if they land on a suitable food such as our skin, lungs or clothing. They fall on the objects in a room, such as equipment and benches, and on our skin and our clothes. There are many on our skin and clothes and in our lungs when we enter a laboratory; some of those in our lungs we breathe out.

When we wash, microorganisms that live in water are left in the moisture that remains on our skin even after we say we have dried it. The same happens when we wash equipment in a laboratory: microorganisms that live in water are left on the equipment afterwards (though, unlike those on our skin, they lack food).

Cultures we make in the laboratory may be contaminated by microorganisms from any of these sources. Unless care is taken, they may also be contaminated by microorganisms in chemicals, including water, used to make the media.

Figure 2.9 shows how the spores and active cells of unwanted microorganisms get everywhere in a laboratory. Using aseptic techniques reduces the risk of contamination of cultures by these unwanted microorganisms.

(b) Sterilisation
Both the equipment and the medium to be used in making a culture must be sterilised to get rid of the unwanted microorganisms on or in them. Sterilisation must be carried out before some experiments and before some industrial processes. 21

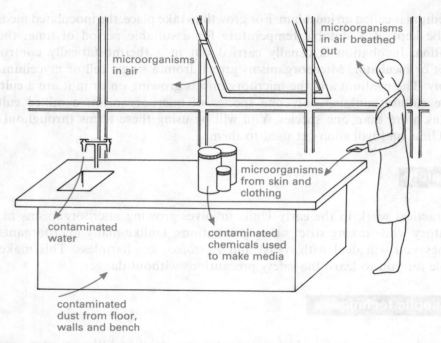

microorganisms in air

microorganisms in air breathed out

microorganisms from skin and clothing

contaminated water

contaminated chemicals used to make media

contaminated dust from floor, walls and bench

Figure 2.9   Sources of contamination in a laboratory

When microorganisms have been used, it is just as important to carry out sterilisation after an experiment or industrial process as before it. Sterilisation must be carried out with care because it can kill human cells. The method of sterilisation will depend on what is being sterilised.

(i) Heat

Nearly all bacteria and fungi are killed by high temperatures (see Section 4.3). The dry heat of a naked flame kills most active cells, but spores are more resistant. To make sure that spores on them are killed, equipment, chemicals (including media) and biological materials should be kept in the dry heat of a hot-air oven at 140–180°C for 1½ to 2 hours plus the time it takes for everything to warm up once inside the oven. Things can be sterilised at these temperatures only if they neither melt nor are destroyed in other ways.

It is usual to sterilise inoculating loops and the necks of glass flasks by flaming them (putting them in a naked flame). Flaming is good enough for most purposes, but it does mean that some of the spores on them may survive. Carcasses and some kinds of equipment no longer needed after an experiment can be burnt: this destroys them and sterilises them at the same time.

Moist heat has a greater effect on cells than dry heat. (This is why you feel hotter in a warm damp atmosphere than in a warm dry one.) Provided they are not damaged by it, apparatus and media can be boiled in water at 100°C for an hour, when only a few bacterial spores will survive.

Pressurised steam is an even better way of killing microorganisms. You have come across pressurised steam if you have seen a pressure cooker, which cooks food faster than an ordinary saucepan or an oven. A pressure cooker boils a small

22

amount of water to make steam, which is pressurised because it is prevented from escaping by a tight-fitting lid and rubber seals. The pressure cooker does not burst because it is made of strong metal and because a release valve in the lid allows steam to escape before the pressure becomes dangerously high.

Pressurised-steam sterilisers, called **autoclaves**, are usually operated at 100 kPa (or 15 lb per in^2), which is enough to kill nearly all heat-resistant spores in 15 minutes. Autoclaves are used to sterilise heat-resistant culture media, glass equipment, towels and most metal items.

Figure 2.10 shows an industrial autoclave. It is like a large pressure cooker laid on its side: the door at the front is its 'lid'. The operator fills it with objects to be sterilised and clamps the door tightly in place. An electric heater boils the small amount of water inside it to make steam, which drives all the air from the autoclave, after which the outlet valve is closed. The steam pressure increases until it reaches a pre-set value. When the pressure has stayed at this value for 15 minutes or however long is necessary, the heater is turned off and the autoclave is allowed to cool down. Only when the pressure inside the autoclave has fallen close to atmospheric pressure can the operator safely open the autoclave and remove the sterile objects.

(ii) Irradiation

All forms of radiation transfer energy, but not all forms can be used for sterilisation. For example, visible light and long-wave radio waves (which your radio converts into sound) are harmless to microorganisms. But short-wave ultraviolet rays,

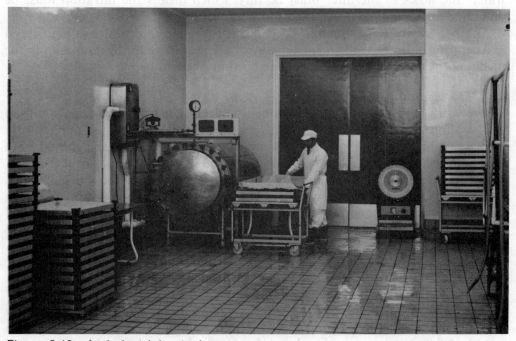

Figure 2.10   An industrial autoclave

23

X-rays and gamma rays (γ rays) are **ionising**. This means they have sufficient energy to damage molecules (see Section 7.5). They can therefore kill microorganisms including spores.

Ultraviolet lamps (making ultraviolet rays) are often used to sterilise the air in laboratories where aseptic work is done. (Because ultraviolet rays can damage the eyes, they must be arranged so that people do not look directly at them.) Some pre-packed plastic Petri dishes and syringes have been sterilised by gamma rays.

(iii) Other methods of reducing contamination

Chemicals used for reducing bacterial contamination include antibiotics, disinfectants and antiseptics. **Antibiotics** are chemicals which can be produced (in small amounts) by some microorganisms and prevent the growth of other microorganisms. Disinfectants are chemicals that kill most microorganisms. They should only be used on non-living surfaces such as work surfaces and clothing. **Antiseptics** are chemicals that either kill microorganisms or slow down their growth and are safe to use on human skin.

Disinfectants are used to sterilise benches in a laboratory. When you are working with microorganisms, you should have some disinfectant ready to use if you accidently spill something on the bench. You must be careful not to get disinfectant on your skin. Antiseptics do not harm your skin, but they do not harm microorganisms very much either, which means they are of little use in a laboratory.

Ultrasonic sound waves (sound waves we cannot hear) are used to kill microorganisms in places it is difficult to reach.

If microorganisms contaminating a culture medium cannot be killed, they must be removed. They can be filtered from the medium much as sand is filtered from water: the holes in the filter must be tiny, which means the filtration must be done under pressure or it will take too long.

Another way of removing microorganisms is to spin the culture at high speed in a centrifuge. Because the microorganisms are heavier than the molecules in the medium, they end up as a pellet at the bottom of the tube.

(iv) The laboratory environment

Before work is begun on cultures, all bench surfaces should have been sterilised with a solution of disinfectant or with an ultraviolet lamp. All equipment and media should have been sterilised. Containers of cultures or media should have been plugged with sterile cotton wool or covered by sterile metal or plastic caps. It is a good idea to light a Bunsen burner at each side of the working area: because it causes an upward draught of warm air away from the bench, it reduces the risk that airborne microorganisms will land in the working area. There should be disinfectant solution near the working area to deal with anything contaminated accidently.

(v) Aseptic transfer of media and microorganisms

You should flame the neck of a container such as a bottle or a tube immediately the plug or cap has been removed and just before it is replaced. You do this by quickly rotating the neck of the container in a Bunsen flame. Figure 2.11 shows a student flaming the neck of a glass tube. Notice how he has hold of the cap of

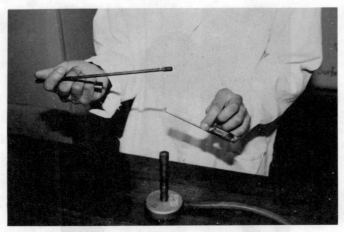

Figure 2.11   Flaming the neck of a tube of medium

this tube in the crook of the little finger of his right hand. The cap itself might become contaminated if he put it down on the bench.

Petri-dish lids should be lifted at a small angle for as short a time as possible (see Figure 2.12).

An inoculum of microorganisms is usually removed from a solid medium with a wire loop. You should flame the loop and cool it before using it. After you have transferred the inoculum, you should flame the loop again before returning it to the bench. You should put the loop slowly into the flame to make sure that the culture's microorganisms do not spray into the air. Figure 2.13 shows how you can sterilise the whole length of a loop by holding it almost vertically in a blue Bunsen flame until it glows.

Liquid inocula can be transferred with a loop but you may prefer to use a pipette. Even though the pipette will already have been sterilised, you should flame it along its whole length and cool it before putting it in the culture. You should not flame it after using it because this would spray the culture's microorganisms into the air.

(vi) Aseptic disposal of contaminated material

All equipment and media must be sterilised before being stored or thrown away. Pipettes must be put in a tall container filled with disinfectant. Wire loops must be flamed. Containers such as tubes, bottles and Petri dishes must be autoclaved (even though plastic Petri dishes will melt and will have to be thrown away).

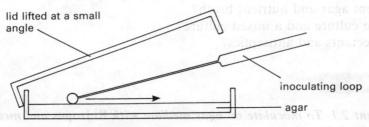

lid lifted at a small angle

inoculating loop

agar

Figure 2.12   Lifting the lid of a Petri dish

25

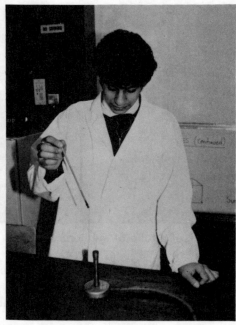

Figure 2.13 Flaming a wire loop

Q 2.1 Some of the following parts are found in yeast cells, some in bacteria and some in both:

cell-surface membrane Y   cell wall YB   chromosome Y   cytoplasm YB
DNA YB   flagellum B   food stores YB   nucleus Y   nuclear envelope Y
vacuole Y

Make a list of these parts. Put a Y against any part that is found in yeast cells and a B against any part that is found in bacteria. Some parts should have both a B and a Y against them.

Q 2.2 Make simple labelled drawings of a yeast cell and of a piece of hypha of *Rhizopus*.
(a) Make a list of the parts found in both fungi.
(b) List as many differences as you can between these two fungi.

Q 2.3 What is the difference between
(a) nutrient agar and nutrient broth?
(b) a pure culture and a mixed culture?
(c) disinfectants and antiseptics?

Practical work

Experiment 2.1 To inoculate an agar medium with **Rhizopus** *and measure the growth of the colony*

26

Note to teachers

Before allowing students to carry out this experiment you should read *Microbiology: An HMI guide for schools and non-advanced further education* (HMSO, 1985).

This experiment lasts at least a week. A practical class is needed for students to pour their own agar plates and inoculate them with *Rhizopus*. The *Rhizopus* cultures should be incubated at 25°C for the duration of the experiment. The students should be allowed access to the laboratory at a certain time each day to measure their *Rhizopus* colonies.

Strict supervision is needed when students open the Petri dish containing the stock culture of *Rhizopus* to ensure that they do not risk releasing spores into the air. You should demonstrate the way to open the dish as little as possible (see Figure 2.12) or you may prefer to inoculate the agar yourself. Students often find it difficult to flame the top of a tube of molten agar: it is worth demonstrating this technique before they begin the experiment.

Pure cultures of *Rhizopus* can be obtained from biological suppliers.

A day or two before the practical class, make enough potato-dextrose agar for the students according to the supplier's instructions. Before autoclaving, pipette 10 cm^3 samples of the liquid agar into heat-resistant test-tubes (one test-tube per student). Either plug each test-tube with non-absorbent cotton wool or cover its top with aluminium foil. Autoclave the test-tubes and their contents at a gauge pressure of 100 kPa (15 lb in^{-2}) for 15 minutes. Store them in a refrigerator.

On the day of the practical put the test-tubes in a boiling-water bath for 10 minutes to melt the agar. Allow the temperature of the water bath to drop only to about 45°C to keep the agar molten until the students use it.

Sodium-chlorate(I) solution deteriorates during storage and should be freshly made before this practical.

This experiment can be done later at the same time as Experiment 7.1.

Materials needed by each student
1 sterile Petri dish
1 inoculating loop
1 Bunsen burner
1 spirit marker
adhesive tape
about 100 cm^3 1% sodium-chlorate(I) solution to deal with spillages
1 plugged or covered test-tube of molten sterile agar in a water bath at 45°C

Materials needed by the class
an incubator at 25°C
hand-washing facilities in the laboratory
an active agar-based culture of *Rhizopus* in a Petri dish

Instructions to students
1. Your Petri dish is sterile. It is important to keep it sterile. Open it only when these instructions say you should. Even then, open it as little as possible, keeping the lid down on one side so that it opens at an angle, and do not 27

breathe on it (see Figure 2.12). Treat the Petri dish containing the culture of *Rhizopus* in the same way: you must avoid contaminating it and you must also avoid releasing spores into the air.

2. Turn your Petri dish upside down on its lid. Use the spirit marker and a ruler to draw two lines at right angles to each other across the middle of the base of the dish. Write *Rhizopus* (the name of the fungus to be grown), your initials and the date near the edge of the dish. Turn the dish the right way up and leave it on the bench.

3. Take a test-tube of molten agar from the water bath. Remove its plug or cover and quickly rotate its top in a blue Bunsen flame to sterilise it (see Figure 2.11). Immediately pour the contents of the test-tube into your Petri dish, lifting the lid at an angie so that only the very top of the test-tube gets over the dish. Immediately replace the lid and leave the dish undisturbed until the agar cools and sets. This will take about 15 minutes.

If you spill any of the molten agar, inform your teacher.

4. When the agar is set, sterilise the inoculating loop (see Figure 2.13) by holding it in a blue Bunsen flame until the wire at its end glows. Take the wire out of the Bunsen flame and let it cool without touching anything.

5. Go to the Petri dish containing the culture of *Rhizopus*. Lift its lid so that you can just get the sterile inoculating loop on to the *Rhizopus*. Use the loop to remove a small piece of the *Rhizopus* from the edge of the colony. Immediately close the Petri dish containing the culture and lift the lid of your own Petri dish just enough to allow you to put the *Rhizopus* on the agar where the two lines you drew cross each other. Immediately withdraw the loop and close the lid.

If you drop any *Rhizopus*, pour some of the sodium-chlorate(I) solution over it and inform your teacher.

6. Put the wire of the inoculating loop back into the Bunsen flame until it glows and then put it down to cool.

7. Stick two short pieces of adhesive tape across the edge of the dish to fix the lid to the base. Do not seal the lid all the way round the dish.

8. Put your dish, base uppermost, in the incubator at 25°C. Wash your hands before leaving the laboratory.

9. At the same time each day for at least a week, return to the laboratory and remove your Petri dish containing your culture from the incubator. Do not open its lid. Put a ruler along one of the lines on the base of the dish and measure the diameter of the growing colony. Do the same along the other

line, find the mean (average) of the two measurements and record it in a table of your own design. Appendix A helps you design a table. You find the mean (average) by adding the two measurements and dividing the sum by 2.

10. At the end of your experiment give the unopened *Rhizopus* culture to your teacher.

Interpretation of results

1. Draw a graph to show the average diameter of your *Rhizopus* colony on the vertical axis against time in days along the horizontal axis. (Appendix A tells you how to draw a graph.)

2. Did the *Rhizopus* grow at a constant rate? Suggest explanations of any differences in rate which appear on your graph (see Section 3.2).

3. Explain why you were told to:
   (a) rotate the top of the tube of agar in a blue Bunsen flame (instruction 3)
   (b) lift the lid of the Petri dish at an angle when pouring the agar (instruction 3)
   (c) hold the wire of the inoculating loop in a blue Bunsen flame until it glowed (instruction 4)
   (d) prevent the inoculating loop touching anything as it cooled (instruction 4)
   (e) avoid fully opening the dish of *Rhizopus* (instruction 5)
   (f) take a piece of *Rhizopus* from the edge of the colony rather than from the middle of it (instruction 5)
   (g) hold the wire of the inoculating loop in a blue Bunsen flame before putting it down after inoculating the agar medium with *Rhizopus* (instruction 6)
   (h) avoid sealing the lid of the Petri dish all the way round (instruction 7).

4. Suggest how you could adapt this experiment to test the hypothesis that temperature affects the growth rate of *Rhizopus*. The adapted experiment is suitable for an Individual Study.

GROWTH IN MICROORGANISMS

3.1 Growth requirements

In order to grow, microorganisms need
○ water
○ suitable food
○ a suitable temperature
○ a suitable pH (suitable acidity or alkalinity)
○ surroundings of a suitable concentration

Different microorganisms have different needs. Some need a good supply of oxygen; others can grow only if oxygen is absent. Yeast needs
○ food in the form of sugar
○ a suitable source of nitrogen such as ammonium ions, urea or amino acids
○ inorganic ions such as phosphates and sulphates
○ minute traces of certain metal ions such as magnesium and calcium
○ minute traces of growth factors such as vitamins

The easy way to grow a culture of microorganisms in the laboratory is to buy a suitable growth medium from a biological supplier and make it up with sterile distilled water according to the maker's instructions. Such a growth medium contains, in the correct proportions, all the needs of the microorganism for which it has been prepared.

(a) Water

All living organisms need water. Water
○ carries dissolved substances into a microorganism
○ allows chemical reactions to take place in solution inside the microorganism
○ is a means of removing dissolved waste products from the microorganism
Water forms the major part of any microorganism. Although an agar medium is solid, it contains plenty of water to meet the needs of microorganisms.

(b) Food

All living organisms need food. They need food as the raw material from which they build cytoplasm, nuclei and walls. They also need food as a source of energy in order to carry out living activities, especially in order to make the complex compounds of their cytoplasm, nuclei and walls. Yeast, for example, needs carbon in a complex form such as sugar but needs other elements, such as nitrogen, phosphorus and sulphur, only as simple inorganic ions. Other microorganisms may need more complex food.

31

(c) Suitable temperature

All organisms grow only within a small temperature range. Somewhere within this range is the **optimum temperature**, the one at which they grow best.

(d) Suitable pH

Microorganisms may need a certain range of acidity or alkalinity. Yeast grows best in slightly acid surroundings: its optimum pH is pH 6.

(e) Surroundings of a suitable concentration

If the concentration of substances outside a microorganism is higher than the concentration of substances inside it, water will pass out of it until the two concentrations are the same. As a result the microorganism may dry out so much that it dies.

3.2 Growth of a yeast colony

Table 3.1 shows the number of yeast cells that developed in a yeast culture from one parent cell during 60 hours.

Figure 3.1 is a graph of these figures with the time in hours which the colony has been growing spaced along the horizontal (x) axis and the number of cells in the colony spaced along the vertical (y) axis. (Appendix A tells you more about graphs.) Look at Figure 3.1: at first the cells do not divide at all; they then multiply quite rapidly, almost doubling themselves every four hours, after which the number in the colony levels off before it finally decreases.

When yeast cells are first added to a liquid growth medium, they do not usually grow and multiply straight away. They first adjust to their new surroundings: this is the **lag** phase of growth. The cells may not have been very active where they came from and they may need to make some substances before then can use the new food supply that now surrounds them. If the cells have come from surroundings like the ones they are now in, the lag phase of growth may be very short or absent altogether.

The lag phase ends when the cells have adjusted and begin to divide. When all the cells in the colony are actively dividing, growth is in the **exponential** (logarithmic) phase: in this phase, one cell divides into two, two cells divide into four, four cells divide into eight, eight into sixteen, and so on. In other words, every cell in the colony is healthy and is dividing. Exponential growth comes to an end when either the supply of an essential food runs out or waste products from living processes start to poison the cells.

Whatever does most to slow down (and eventually stop) growth is called the **limiting factor**: it may be some food needed only in small amounts or some waste

Table 3.1

Hours	0	4	8	12	16	20	24	28	32	36	40	44	48	52	56	60
Number of yeast cells	1	1	2	5	8	15	29	52	86	116	134	142	143	142	130	85

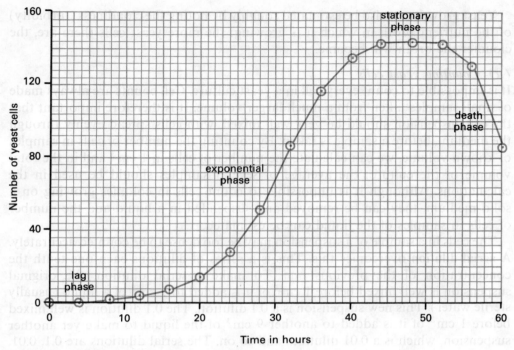

Figure 3.1   Growth of a yeast colony

product produced only in small amounts. (Whatever most interferes with any trend is the limiting factor.)

At the end of the exponential phase, growth slows down: the number of cells in the colony still increases but the rate at which they increase is reduced. Soon the number of new cells being formed is exactly balanced by the number of cells that are dying: this is the **stationary** phase of growth when the number of cells in the colony does not change even though the cells themselves do change.

The stationary phase is followed by a decrease in the number of cells because the new cells are too few to replace those that are dying. Eventually no new cells are formed and the number of cells in the colony decreases rapidly as the existing cells die: this is the **death** phase.

3.3 Measuring growth

In Experiment 2.1 you measured growth in *Rhizopus* by measuring the diameter of a mycelium every day for a week. An organism grows when it makes more of itself: when it makes more cytoplasm, nuclei and wall. This is what *Rhizopus* was doing as its hyphae spread outwards across the agar.

We do not measure the growth of yeast by measuring the diameter of the colony. For one reason, yeast is a single-celled organism, not a mycelium, and its cells are separated. For another, yeast is usually grown not on a solid medium but in a nutrient broth. If we want to measure the growth of yeast, we count the number of cells it produces in a certain time.

Yeast growth can also be estimated roughly by noting the cloudiness (turbidity) of the nutrient broth in which it is growing: the more yeast cells there are, the cloudier the broth will become.

(a) Counting yeast cells

It is impossible to count every cell in a yeast culture. Cell counts are always made of small samples taken from a much larger volume. It is therefore important that the cell suspension should be very well mixed, with cells spread evenly through the medium, before a sample is taken. By counting the number of cells in samples of known volume, we can calculate the number of cells in 1 cm^3 and in the total volume of the culture: the average of several samples should be used in the calculations. Although it is impossible to count cells in colonies growing on a solid medium, they can be removed and suspended in a liquid and the number of cells in samples of the liquid can be estimated.

The cells in a sample of a suspension may be too dense to be counted accurately. A **serial dilution** overcomes this. This is a series of dilutions each one tenth the concentration of the previous one. To make a serial dilution, the original suspension is well mixed before 1 cm^3 of it is added to 9 cm^3 of a liquid (usually sterile water). This new suspension is a 0.1 dilution. The 0.1 dilution is well mixed before 1 cm^3 of it is added to another 9 cm^3 of the liquid to make yet another suspension, which is a 0.01 dilution, and so on. The serial dilutions are 0.1, 0.01, 0.001, 0.0001 and so on. We say that the 0.1 suspension has a **dilution factor** of 10, the 0.01 suspension has a dilution factor of 100, and so on. All the dilutions are looked at under a microscope to see which is most suitable for counting.

A **total count** estimates the number of cells, living or dead, in a culture. A **viable count** estimates only the number of living cells in a culture. We cannot tell by looking at it if a yeast cell is living or not. If we want a viable cell count, we take samples and count them only during the **exponential** phase of growth. In this phase we can assume that all the cells are living and growing.

In Figure 3.2 you can see a side view of a special **counting chamber** on a microscope slide. Because this type of counting chamber is often used to count red blood cells, it is called a **haemocytometer**. (*Haima* is the Greek for *blood*.) Figure 3.3 is a top view of the grid on the central platform of the haemocytometer slide. When a coverslip is correctly placed across the chamber, the depth of fluid over the grid and central platform is exactly 0.1 mm.

In Figure 3.3 you can see that the grid is a 1 mm square divided by triple lines into 25 squares of side 0.2 mm and that each of these squares is divided by single

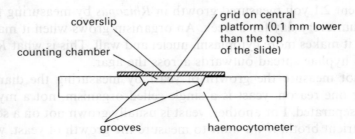

34   Figure 3.2   Side view of a haemocytometer slide and coverslip

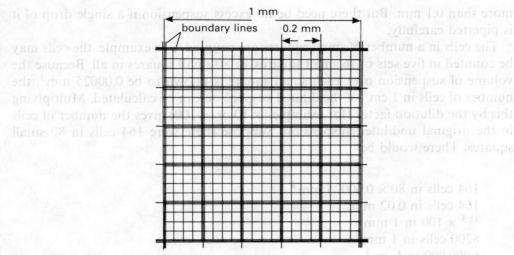

Figure 3.3 Grid on a haemocytometer slide

lines into 16 smaller squares of side 0.05 mm. The area of the smallest squares is therefore 0.0025 mm^2 (0.05 mm × 0.05 mm). With the coverslip in place and a depth of 0.1 mm above the grid, a volume of fluid of 0.00025 mm^3 (0.0025 mm^2 × 0.1 mm) is trapped above each of the smallest squares.

The coverslip is first slid into position above the chamber and pushed against the haemocytometer slide until rainbow-like lines (called **Newton's rings**) appear on the outer side of the two grooves as shown in Figure 3.4. This ensures that the gap above the grid is exactly 0.1 mm. Dilute yeast suspension, from one drop pipetted on to the central platform at the edge of the coverslip, seeps into the chamber and covers the counting grid.

The suspension is given several minutes for the cells to settle and stop moving before it is looked at under high power of a microscope. Because the distance from the central platform to the coverslip is only 0.1 mm, the suspension is sure to fill the space between them. The grooves shown in both Figures 3.2 and 3.4 allow any excess suspension to get out of the counting chamber without forcing up the coverslip and making the distance between it and the central platform

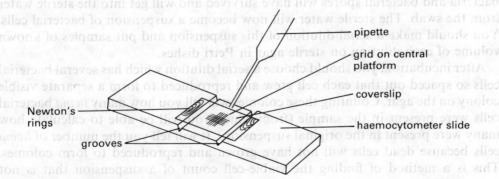

Figure 3.4 Haemocytometer slide

35

more than 0.1 mm. But there need be no excess suspension if a single drop of it is pipetted carefully.

The cells in a number of small squares are counted. For example, the cells may be counted in five sets of 16 small squares, or 80 small squares in all. Because the volume of suspension over each small square is known to be 0.00025 mm^3, the number of cells in 1 cm^3 of the diluted suspension can be calculated. Multiplying this by the dilution factor (for example, $\times 10$ or $\times 100$) gives the number of cells in the original undiluted suspension. Suppose there were 164 cells in 80 small squares. There would be

164 cells in 80×0.00025 mm^3
164 cells in 0.02 mm^3
$\frac{164}{2} \times 100$ in 1 mm^3
8200 cells in 1 mm^3
8 200 000 in 1 cm^3

If there was a dilution factor of $\times 10$, the number of cells in 1 cm^3 of the undiluted suspension would be 82 000 000.

Microbiologists usually decide which squares they will count before looking down the microscope. In this way they avoid the temptation to pick certain squares because they look special. They count not only all cells lying within a square but also all those touching and overlapping the top and left-hand sides of the square. They do not count those that touch or overlap the bottom and right-hand sides because these would be counted in neighbouring squares. There is no reason why this system of counting should be used rather than one involving different sides of the square. This just happens to be the system that microbiologists have agreed to use.

Industrial and medical laboratories use a **Coulter counter** to estimate the number of single-celled organisms in a suspension. A Coulter counter measures a change in voltage as cells pass through an electrical field: cells decrease the voltage because their cell-surface membranes contain lipid, which is a poor conductor of electricity.

You wish to test the disinfectant used in your laboratory. You should use a sterile swab to wipe a surface where disinfectant has been used. Having done that, you put the swab in sterile water. No matter how good the disinfectant, some bacteria and bacterial spores will have survived and will get into the sterile water from the swab. The sterile water will now become a suspension of bacterial cells. You should make a serial dilution of this suspension and put samples of known volume of each dilution on sterile agar in Petri dishes.

After incubation, you should choose a serial dilution which has several bacterial cells so spaced out that each cell grew and reproduced to form a separate visible colony on the agar. Counting these colonies will tell you how many *living* bacterial cells were present in the sample (from which you will be able to calculate how many were present in the original suspension). It will tell you the number of *living* cells because dead cells will not have grown and reproduced to form colonies. This is a method of finding the viable-cell count of a suspension that is not necessarily in the exponential phase. The method is shown in Figure 3.5.

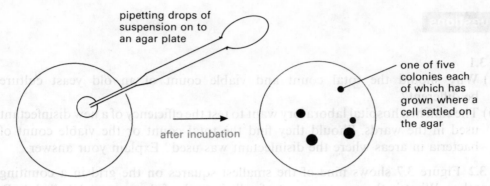

one of five colonies each of which has grown where a cell settled on the agar

pipetting drops of suspension on to an agar plate

after incubation

Figure 3.5   A viable-cell count

(b) Turbidity measurements

Yeast cells absorb and reflect light. The more cells there are in a yeast culture, the less light will pass through it and the greater will be its **optical density** (cloudiness or **turbidity**). As the yeast cells multiply, so the optical density of the culture increases (it becomes more cloudy or turbid). Measuring the optical density of a yeast culture is an alternative to counting yeast cells: it is an indirect way of measuring yeast growth.

Figure 3.6 shows the parts of a **colorimeter** used to measure the amount of light passed from a lamp through a yeast culture in a plastic container. An electrical photocell registers the amount of light that has passed through on a scale: the less the amount of light, the greater the optical density. An orange filter is usually placed between the light source and the yeast culture; other colours would be used with cultures other than yeast.

To be able to read off the number of yeast cells in a culture for any optical density, the colorometer must be calibrated against cultures with known yeast numbers. (Something is calibrated if measurements can be made with it. A ruler is usually calibrated in millimetres and tenths of an inch.)

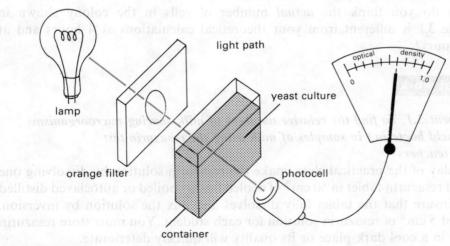

light path

optical   density
0   1.0

lamp

yeast culture

orange filter

photocell

container

Figure 3.6   Parts of a colorimeter

37

Q 3.1

(a) Why should the total count and viable count of an old yeast culture be different?

(b) The staff in a hospital laboratory want to test the efficiency of a new disinfectant used in the wards. Should they find the total count or the viable count of bacteria in areas where the disinfectant was used? Explain your answer.

Q 3.2 Figure 3.7 shows nine of the smallest squares on the grid in a counting chamber. What is the correct count of cells in each of the squares labelled A, B, C and D?

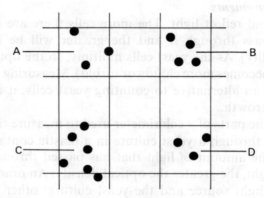

Figure 3.7   Part of a counting-chamber grid

Q 3.3 Assume that yeast doubles itself every four hours and that you start with one yeast cell at 0 hours.

(a) Make a table like Table 3.1 showing the *theoretical* growth in the number of yeast cells during a period of 60 hours.

(b) Why do you think the *actual* number of cells in the colony shown in Table 3.1 is different from your theoretical calculations at 4 hours and at 44 hours?

Practical work

Experiment 3.1 To find the relative numbers of milk-souring microorganisms (lactic-acid bacteria) in samples of milk using the resazurin test

Note to teachers

On the day of the practical class, make the resazurin solution by dissolving one standard resazurin tablet in 50 cm^3 of cooled freshly boiled or autoclaved distilled water. Ensure that the tablet fully dissolves and mix the solution by inversion. You need 5 cm^3 of resazurin solution for each student. You must store resazurin solution in a cool dark place or its quality will quickly deteriorate.

Collect pasteurised milk supplied on each of the two days before this practical class and also on the day of the practical class. Collect enough for each student to have 15 cm^3 of milk supplied on each of the three days. Unless the weather is particularly warm, do not store it in a refrigerator. Collect also enough UHT milk for each student to have 15 cm^3: UHT milk can be bought in capped bottles and in cartons.

Students may need help in identifying the colours in instruction 7.

Note to students

When milk is delivered to dairies, it must be checked to make sure it is fit to drink. One test used by dairies is called the 'platform rejection test', because it is carried out on the delivery platforms where churns of milk are unloaded from wagons. Milk below a certain standard is rejected and sent back to the producer.

Dairy technicians use aseptic techniques when they carry out the platform rejection test. This experiment is based on the platform rejection test, but to keep it simple you will not use aseptic techniques.

Materials needed by each student
1 10 cm^3 pipette fitted with a pipette filler
1 1 cm^3 pipette fitted with a pipette filler (or the same pipette filler can be used)
1 test-tube rack with 4 test-tubes fitted with rubber bungs
1 stopclock
1 spirit marker
5 cm^3 fresh resazurin solution
15 cm^3 clearly labelled 2-day-old pasteurised milk
15 cm^3 clearly labelled 1-day-old pasteurised milk
15 cm^3 clearly labelled fresh pasteurised milk
15 cm^3 clearly labelled freshly opened UHT milk

Materials needed by the class
thermostatically controlled water bath at 37.5°C ± 0.5°C

Instructions to students
1. Read through all these instructions before starting and be ready to carry out instructions 5 to 8 fairly quickly.
2. Copy the table.

Tube	Contents	Colour of contents at start	Colour of contents after ten minutes
A	2-day-old milk		
B	1-day-old milk		
C	fresh milk		
D	UHT milk		

3. Use the spirit marker to label four test-tubes A to D.
4. Shake each milk sample before using it. Pipette 10 cm^3 of each milk sample 39

into the test-tube indicated by the table: for example, pipette 10 cm^3 of 2-day-old milk into test-tube A. Clean the pipette out with water between pipetting the different milk samples.

5. Add 1 cm^3 of resazurin solution to each test-tube of milk and start the stopclock.
6. Replace the rubber bungs and turn each test tube upside down twice to mix its contents. Keep the test-tubes sealed with the rubber bungs.
7. Record the colour of the contents of each test-tube in your table using colours in the list below.

> violet
> purple
> pink
> faint pink
> white

8. Put all the test-tubes in the water bath.
9. After ten minutes, remove the test-tubes from the water bath and record the colours of their contents in your table. Again use colours in the list.

Interpretation of results

1. Suggest why you were told to carry out instructions 5 to 8 fairly quickly (instruction 1).
2. Why do you think you were told to keep the test-tubes sealed with the rubber bungs (instruction 6)?
3. Resazurin is a dye which is affected by milk-souring microorganisms. When added to fresh milk, it is violet, but it may change to pink and finally to colourless depending on the number and rate of growth of milk-souring microorganisms. The colours in instruction 7 show the range you would expect from milk with few souring microorganisms (at the top of the list) to milk with many souring microorganisms (at the bottom of the list).

    Using this information and the results of your experiment, suggest which sample or samples might pass the 'platform rejection test'. Give a reason for your answer.
4. What do you think happens to milk during pasteurisation? Do your results show that all milk-souring microorganisms are killed by pasteurisation? If not, do you think the remaining milk-souring microorganisms are harmful to our health? Give a reason for your answer.
5. 'UHT' stands for 'ultra-high temperature'.
    (a) Do your results suggest that this treatment has killed all microorganisms in the milk?
    (b) Suggest why cartons of UHT milk have a 'best by' date stamped on them.
6. Why is it important that dairy technicians use aseptic techniques when carrying out the platform rejection test? Suggest what aseptic techniques they might use.

ENZYMES

4.1 Introduction

A **cell** is the smallest unit of living matter. Inside a cell take place the chemical reactions that are the cell's living activities. These chemical reactions inside cells, which are known as **metabolism** or **metabolic activities**, either release or use energy inside the cell.

Metabolic activities are the same chemical reactions that can be made to take place in a test-tube, yet they take place inside a tiny living cell. In yeast, all the chemical reactions needed to keep yeast alive take place inside the one cell. In large organisms different cells may concentrate on different chemical reactions, but all the cells can both release and use energy.

Metabolic activities are complicated. Though a particular reaction could take place in a test-tube, it might be very slow unless the temperature and pressure were high. Living cells avoid the need for high temperatures and pressures and carry out their metabolic activities at great speed by the use of special chemicals called **enzymes**.

4.2 Enzymes

If you put some hydrogen-peroxide solution in a flask, nothing seems to happen. Even after half an hour nothing will seem to have happened. If you add some manganese(IV) oxide, almost immediately bubbles are given off. You may think there is nothing odd about this. What has happened, it may seem obvious, is that there has been the usual kind of chemical reaction between the manganese(IV) oxide and the hydrogen peroxide.

In fact there has not. If you collect all the gas given off, you will find that it is oxygen. If you study what is left in the flask, you will find that it is water and manganese(IV) oxide. What is more, you will find the same amount of manganese(IV) oxide as you put in. Manganese(IV) oxide has not been used in the chemical reaction. This chemical reaction is shown in words as

$$hydrogen\ peroxide \rightarrow water + oxygen$$

and in chemical symbols as

$$2H_2O_2 \rightarrow 2H_2O + O_2$$

Yet it is obvious that manganese(IV) oxide has had an effect on the chemical reaction in which hydrogen peroxide changed into water and oxygen. It seems that manganese(IV) oxide has caused the chemical reaction. But this is not what 41 has happened either. The truth is that, even before you added the manganese(IV) oxide, the hydrogen peroxide was changing into water and oxygen, but so slowly that you could not see any difference. The effect of the manganese(IV) oxide was to speed up a chemical reaction that was taking place anyway.

A substance that speeds up a chemical reaction is called a **catalyst**.

Now suppose that, instead of manganese(IV) oxide, you add small pieces of liver to hydrogen peroxide in a flask. Again almost immediately bubbles will be given off. There is no manganese(IV) oxide in liver, but there is a biological catalyst that also speeds up the change of hydrogen peroxide (a cell poison) into water and oxygen. This biological catalyst is called **catalase**. If you add yeast cells, instead of liver, to hydrogen peroxide, it too will quickly get rid of the poisonous hydrogen peroxide by speeding up its change into water and oxygen. Most cells contain catalase and are able to speed up the removal of hydrogen peroxide.

Heating the hydrogen peroxide would also have changed it more quickly into water and oxygen. Manganese(IV) oxide, liver and yeast speed up the reaction without extra heat. Inside any living cell there are at least a thousand different catalysts allowing chemical reactions to take place quickly without extra heat. Biological catalysts are called **enzymes**. Enzymes are made in the cytoplasm of a living cell on the instructions of the DNA in the nucleus.

Most non-biological catalysts speed up a number of chemical reactions. Enzymes are different. Each enzyme speeds up only one chemical reaction or one type of chemical reaction. We say each enzyme is **specific** to one chemical reaction or type of reaction. This is why so many different enzymes are needed in any living cell. In a living cell there may be long series of chemical reactions every one of which needs its own enzyme.

(a) How an enzyme works

The smallest particle of an enzyme, or indeed of any substance, which can exist by itself is a **molecule**. Most enzymes are complex molecules made up of long chains of chemicals called **amino acids**. These chains are rolled up, like a loose ball of string, with cross links made from one part of the chain to another to hold the ball in a stable shape. A long chain of amino acids forms a substance called a **protein** (see Section 15.4). When the amino-acid chain is in the form of a ball, it is a **globular protein**. Such enzyme molecules are so complex that even the smallest enzyme molecule is about 6000 times heavier than a molecule of hydrogen. The largest enzyme molecules are over a million times heavier. (Nearly all enzymes are proteins. It was not until the 1980s that an enzyme was discovered which is not a protein.)

Most enzymes are needed only in very small amounts. No one has seen how an enzyme affects the speed of a chemical reaction but there is a **hypothesis** (an unproved theory) which at least explains why each different enzyme affects the speed of only one chemical reaction or type of reaction.

It is believed that every enzyme has molecules of a shape different from those of all other enzymes. The substance or substances involved in a chemical reaction in a cell must be able to fit into and be surrounded by the surface grooves of an enzyme. An enzyme's surface grooves, where substances are enclosed, are called the **active site**. Only substances with exactly the right-shaped molecules will match

42

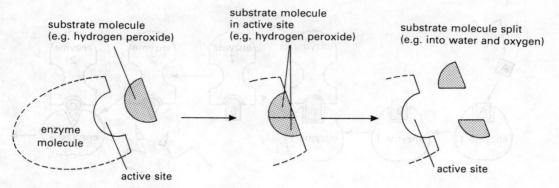

Figure 4.1   Enzyme action: substrate split

the shape of the active site on the enzyme. This is why each enzyme is specific to one chemical reaction or type of reaction.

Figure 4.1 shows how an enzyme can help split a molecule of hydrogen peroxide into water and oxygen. A substance on which an enzyme acts is called a **substrate**. In our example a molecule of the substrate hydrogen peroxide fits into the active site of the enzyme catalase. In some way we do not understand this makes the reaction easier and so speeds it up. When the reaction is complete, water and oxygen are released out of the active site of the enzyme and another molecule of hydrogen peroxide takes their place.

The slowest enzymes take over one second to catalyse one substrate molecule but most enzymes act quickly: catalase can catalyse five million molecules of hydrogen peroxide in a minute.

Enzymes speed up all sorts of different reactions inside a cell. Splitting a molecule is one kind; other kinds include joining molecules together, shown in Figure 4.2, and changing the shape of a molecule, shown in some of the reactions in Figure 4.3.

Many of the metabolic activities that take place inside a cell are a series of chemical reactions each of which needs a different enzyme. **Respiration**, the release of energy inside a cell (see Unit 5), is a series of reactions which needs about 70 different enzymes. One hypothesis to explain how such a series of reactions works

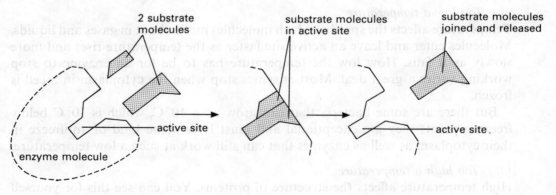

Figure 4.2   Enzyme action: substrates joined

43

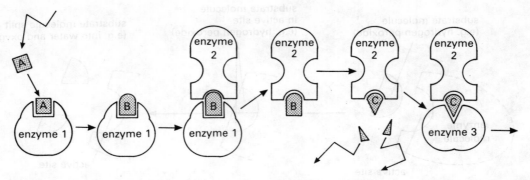

Figure 4.3   Enzymes involved in a series of chemical reactions

is that one enzyme accepts a substrate, changes it in some way and passes it on
to the next enzyme, which changes it and passes it on to the next enzyme, and
so on. This is shown in Figure 4.3.

4.3 Enzyme sensitivity

During one minute catalase can catalyse at least five million molecules of substrate.
To work at its fastest, catalase needs **optimum** conditions of, for example,
temperature and pH. The optimum is the best when the best is neither the minimum
nor the maximum. The best time for an athlete in a race is the fastest and is *not*
an optimum. The best temperature for running the race is when it is neither too
hot nor too cold. The best temperature, when the athlete can run fastest, is the
optimum temperature.

(a) Temperature

Most enzymes work best at a moderate temperature. Most enzymes in the human
body work best at about 37°C. Even a few degrees either way may decrease the
efficiency of the enzyme. There is a **minimum** temperature below which the enzyme
will have no effect on the chemical reaction and a **maximum** temperature above
which it will also have no effect on the chemical reaction. The temperatures
between the minimum and maximum are the **range** over which the enzyme works.

(i) Too low a temperature

Temperature affects the speed at which molecules move about in gases and liquids.
Molecules enter and leave an active site faster as the temperature rises and more
slowly as it falls. How low the temperature has to be for an enzyme to stop
working varies a great deal. Most enzymes stop when the cytoplasm in a cell is
frozen.

But there are some bacteria that can grow at −10°C, which is 10°C below
freezing point: they are exceptional and must have some kind of antifreeze in
their cytoplasm as well as enzymes that can still work at such a low temperature.

(ii) Too high a temperature

High temperature affects the structure of proteins. You can see this for yourself
when an egg is cooking. The white of the egg, the albumen, is a globular protein

44

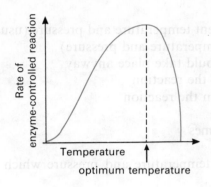

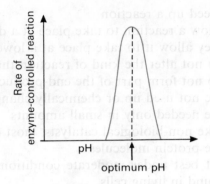

Figure 4.4   Effect of temperature on the rate of an enzyme-controlled reaction

Figure 4.5   Effect of pH on the rate of an enzyme-controlled reaction

which, as it heats up, changes from a colourless jelly-like substance to a semi-solid state and finally to the solid white of a hard-boiled egg. What has happened is that the protein molecule has lost its highly organised and complex structure and has become **denatured**. All proteins become denatured if heated beyond a certain temperature, and nearly all enzymes are proteins. Therefore, as the temperature rises, the enzyme becomes denatured, the shape of its active site is destroyed and it no longer works.

There are, however, bacteria that grow and reproduce at a temperature of 250°C. They live at a depth of about 2500 m, under enormous pressure, on the floor of the Pacific Ocean. These bacteria must be very unusual to be able to work at that temperature and under such pressure; we know very little about them.

Figure 4.4 shows the usual relation between temperature and the rate of an enzyme-controlled reaction. Below the optimum temperature the rate of the reaction slowly falls as the substrate molecules slow down. Above the optimum temperature the rate of the reaction falls quickly as the enzyme molecules are denatured.

(b) pH

Just as enzymes work only within a narrow temperature range, so also they work only within a narrow pH range. For most enzymes the optimum pH is neutral (pH 7) or something close to it. But some exceptional enzymes work in highly acid conditions (near pH 1) and others work in highly alkaline conditions (near pH 14).

An unsuitable pH makes the active site of an enzyme unable to accept the substrate molecules. It does so because, like too high a temperature, it affects the structure of the enzyme molecule and so changes the shape of the active site. Figure 4.5 shows that the rate of an enzyme-controlled reaction slows rapidly and more or less equally either side of the optimum pH.

4.4 Enzymes and chemical catalysts

Like non-biological catalysts, enzymes

- speed up a reaction
- allow a reaction to take place at a different temperature and pressure (usually they allow it to take place at a lower temperature and pressure)
- do not alter the kind of reaction that would take place anyway
- do not form part of the end-products of the reaction
- are not used up or chemically changed in the reaction
- are needed only in small amounts

Unlike non-biological catalysts, most enzymes
- are protein molecules
- act best under moderate conditions of temperature and pressure which are found in living cells
- are highly sensitive to temperature and pH and act well only within narrow ranges of temperature and pH
- are specific, speeding up only one reaction or type of reaction

4.5 Enzymes *in vitro*

A hundred and fifty years ago scientists reliased that enzymes would work outside cells as well as inside them. They used them *in vitro*, which means separated from the living organism that made them. (Literally *in vitro* means *in glass*, as in a test-tube.) In fact humans had used enzymes *in vitro* for centuries when they made cheese, but they had not realised it.

Even today few enzymes are used *in vitro*. This is because in a laboratory it is usually difficult and expensive to remove enzymes from the living cells that make them. Fortunately some enzymes are removed from cells by natural processes. All we need do then is collect them. The enzymes that are removed from cells naturally are those concerned with digestion.

Digestion is the breakdown of food, often from a solid form, into a form which will dissolve in water, which has small molecules and which will pass easily into an organism's cells. Digestive enzymes are made in the cytoplasm of a cell and secreted (poured out) on to the food outside it in order to digest it. Bacteria and the hyphae of fungi digest food outside their bodies.

Bacteria and the hyphae of fungi secrete enzymes on to their food, which is digested into a simpler **soluble** (dissolvable) form. Provided that water is also present, the soluble food dissolves, is absorbed through the wall and cell-surface membrane which surround the organism, and passes into the cytoplasm, where it can be used. This process of **external** or **extracellular digestion** and **absorption** in bacteria and fungi is shown in Figure 4.6. Enzymes produced in this way are the easiest for us to collect.

Mammals such as humans and cattle digest food inside their stomachs and intestines, but the process is the same: the cells secrete digestive enzymes on to the food outside them. One of the digestive enzymes secreted by the cells lining the stomach of a calf is **rennin**. Secreted into the calf's stomach, rennin separates the solid protein from the liquid whey in the milk that it drinks from its mother, the cow. For hundreds of years cheese-making has made use of rennin obtained

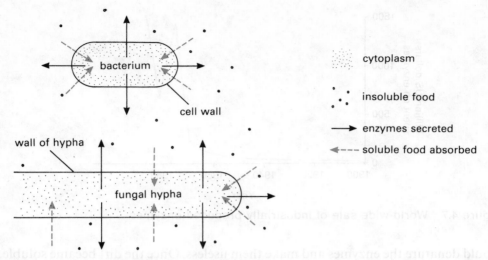

cytoplasm

insoluble food

enzymes secreted

soluble food absorbed

bacterium

cell wall

wall of hypha

fungal hypha

Figure 4.6   External digestion and absorption

from the stomachs of dead calves: rennin separates the solid protein for cheese-making from the liquid whey.

Many farm cheeses are still made from rennin taken from the stomachs of dead calves, but rennin can now be obtained from fungi and even from genetically engineered bacteria (see Section 25.4). Genetic engineering will in future give us many enzymes at a reasonable cost. For the time being we shall continue to collect those produced naturally, such as rennin, and others produced by bacteria and fungi.

(a) Economic advantages of enzymes

When we get enzymes to work for us in factories, costly and complicated chemical processes can be carried out more efficiently and more cheaply. There is no need for high temperatures and pressures. Several enzymes, especially those secreted in large amounts during the digestion processes of bacteria and fungi, have been collected and used for many years. These are still the most commonly used enzymes because they are produced and secreted by the organisms in fairly large amounts and can be removed at a reasonable cost from the broth in which the organisms are cultured. The brewing, baking and textile industries have used starch-digesting enzymes for some time. The dramatic increase in the manufacture and sale of industrially produced enzymes came in the mid 1960s.

Figure 4.7 shows how the world-wide sale of industrially produced enzymes increased gradually from zero at the beginning of the century to the mid-sixties, soared through the late sixties and early seventies and then fell suddenly before taking off again in the early eighties. What was all this about?

The answer is **biological washing powders**. Protein-digesting enzymes were produced in vast amounts by bacteria cultured in a nitrogen-rich medium. These bacteria were not cultured in Petri dishes but in industrial **fermenters**, huge containers designed for culturing microorganisms on a large scale (see Section 7.6). Added to washing powders, these enzymes digested protein stains such as blood (which had always been difficult to remove from clothes) at moderate temperatures. The washing had to be done at moderate temperatures because high temperatures   47

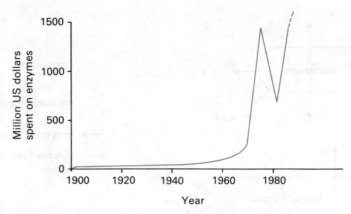

Figure 4.7   World-wide sale of industrially produced enzymes

would denature the enzymes and make them useless. Once the dirt became soluble, it dissolved in the water in which the clothes were washed and was removed when they were rinsed. Biological washing powders were successful.

So what happened in the mid-seventies to make the sales of biological washing powders fall? Workers producing them and consumers using them complained of allergic reactions such as skin rashes, 'hay fever', eczema and asthma. Sales continued to fall until the manufacturers found the solution: they enclosed the enzymes in a harmless coating and introduced further safety measures in their factories.

The coating prevented both workers and consumers from touching the enzymes while the powder was dry. Even when consumers put the biological washing powder in warm water, it took between quarter and half an hour for the coating to dissolve and allow the enzymes to do their work. Consumers had usually taken their hands out of the water long before this. They no longer got skin rashes or eczema from touching the enzymes. Nor, because the enzymes were freed only in water, did they get hay fever and asthma.

Enzymes were first removed from inside cells in the late 1920s. To remove the enzyme undamaged was an expensive and difficult process. When the enzyme was used in a chemical reaction, such as changing starch to sugar, there was another expensive and difficult process to separate the enzyme from the end-product. Industry wants to use an enzyme to catalyse the same reaction again and again. The expense of separating enzymes from end-products usually meant it was not economic to use them again. For a long time, therefore, enzymes were not used industrially on any scale.

The breakthrough came in the late 1970s when it was discovered how to fix enzymes to the solid surfaces of plastics, cellulose fibres and even glass beads. When an enzyme has been used, for example to change starch to sugar, the sugar solution can be drained away leaving the enzyme stuck to its support ready for the next batch of starch. Enzymes used in this way are called **immobilised enzymes**. The manufacturer uses the same immobilised enzyme again and again. New uses for immobilised enzymes are discovered all the time: we are only at the beginning of **enzyme technology**.

48

Though we say we are living in a technological age, technology as such is not new. **Technology** is any deliberate use of materials or energy to meet human needs. Early humans' use of sharpened stones for cutting up animals or to fight with was technology. Great technological advances include the inventions of the wheel, the steam-engine and the internal-combustion engine (used in cars). Enzyme technology is one form of **biotechnology** (see Section 6.1), the use of biological agents to meet human needs.

Questions

Q 4.1
(a) What happens during the removal of a blood stain by a biological washing powder?
(b) Why may clothes get cleaner if they are left to soak in a biological washing powder before they are washed?

Q 4.2 Biological washing powders cost about the same as ordinary washing powders and are used in roughly the same quantities. Yet it is claimed that it is more economical to use biological washing powders than ordinary ones. Do you think this is true? Explain your answer.

Q 4.3 Why must the temperature and pH be carefully controlled during the manufacture of a biological washing powder?

Practical work

Experimental 4.1 To test the hypothesis that catalase (in yeast) breaks down hydrogen peroxide
Note to teachers
About 30 minutes before the practical class starts, prepare 15 cm^3 per student of the yeast–glucose mixture. To prepare 250 cm^3, add 8 g of glucose and 6 g of dried baker's yeast to 250 cm^3 of water at 45°C. Stir the mixture and leave it in a warm incubator or drying cabinet until the practical class starts.

Materials needed by each student
1 250 cm^3 beaker
1 5 cm^3 graduated pipette fitted with a pipette filler
1 boiling tube
1 test-tube rack with 2 test tubes
1 stopclock
1 Bunsen burner, tripod and gauze
1 ruler graduated in mm
1 spirit marker
1 pair safety spectacles
1 pair gloves
15 cm^3 5-volume hydrogen-peroxide solution
15 cm^3 yeast–glucose mixture

49

Instructions to students

1. Half fill the beaker with tap water and make a water bath by boiling the water using the Bunsen burner, tripod and gauze. Put about half the yeast–glucose mixture into the boiling tube and leave it in the boiling water bath for at least 5 minutes while you continue carrying out the instructions.
2. Copy the table.

	Height of froth in mm	
Time in minutes	Tube A (with the boiled yeast–glucose mixture)	Tube B (with the unboiled yeast–glucose mixture)
0	0	0
2		
4		
6		
8		
10		

3. Use the spirit marker to label one test tube A and the other test tube B. Put the test tubes into the test-tube rack.
4. **Wearing safety spectacles and gloves**, pipette 5 cm^3 of hydrogen-peroxide solution into tube A and 5 cm^3 of hydrogen-peroxide solution into tube B.
5. Cool the boiled yeast–glucose mixture. Clean the pipette and use it to put 5 cm^3 of the boiled and cooled mixture into tube A.
6. Clean the pipette, use it to put 5 cm^3 of the unboiled yeast–glucose mixture into tube B, and immediately start the stopclock.
7. At 2-minute intervals, use the ruler to measure the height of the froth (in mm) which may have appeared above the liquid in each tube. Enter the heights you measure in your table.

Interpretation of results

1. Draw a graph of your results with 'Time in minutes' on the horizontal axis and 'Height of froth in mm' on the vertical axis. Plot two curves, one for the boiled and the other for the unboiled yeast–glucose mixture. Appendix A tells you how to draw a graph.
2. Was the increase in the height of the froth in tube B the same for each 2-minute interval? If not, can you suggest why?
3. Explain the difference between the curves for the boiled and unboiled yeast–glucose mixtures.
4. The boiled yeast–glucose mixture was used as a 'control'. What does this mean? Why was boiled yeast–glucose mixture chosen as a control?
5. How do you know that hydrogen peroxide has been broken down in tube B?

50

6. Write a word equation to show what happened to the hydrogen peroxide in tube B.
7. Does this experiment in fact prove that it was the catalase in yeast which broke down the hydrogen peroxide? Explain your answer.

Experiment 4.2 *To test the hypothesis that catalase is sensitive to temperature and to pH*
Note to teachers
The 'Instructions to students' assume that each of them will design the two experiments and hand in outlines of the methods and lists of practical requirements.

Students must do Experiment 4.1 before this one. They will need the apparatus and reagents used in Experiment 4.1 as well as those used to vary temperature and pH. You should tell students what equipment and reagents are available in the laboratory to control temperature and pH.

Instructions to students
1. Design two experiments; one to test the hypothesis that the enzyme catalase is affected by temperature and the other to test the hypothesis that the enzyme catalase is affected by pH. Some clues to help you design your experiments are given below. If after reading them and thinking about them you do not know how to start, ask your teacher for help.
2. Make a list of the steps you will take during each of your experiments. Show it to your teacher, who will check that your methods work.
3. Make a list of the apparatus and solutions you will need to do your experiments. Give it to your teacher in advance so that everything you need can be got ready.
4. When you have done each of your experiments, write a report telling what you did, what the results were and what these show about the hypothesis you are testing. Appendix B helps you write a report.

Clues to help you design your experiments
1. There are several ways of testing each hypothesis. You need not worry if your way is different from that of other students in your class.
2. Do not try to test the effects of temperature and pH in a single experiment. Design one experiment to test the effect of temperature and another to test the effect of pH.
3. Remind yourself of how you measured the effect of catalase on hydrogen peroxide in Experiment 4.1. In both the experiments you are designing you can use the same equipment and solutions as you used in Experiment 4.1.
4. If catalase is affected by temperature, the height of froth produced from the hydrogen peroxide may be different at different temperatures. If catalase is affected by pH, the height of froth produced from the hydrogen peroxide may be different at different pH values. How can you adapt Experiment 4.1 to find these heights?
5. At how many different temperatures will you measure the height of froth produced? How will you keep each of these temperatures constant while you carry out the experiment?

51

6. At how many different pH values will you measure the height of froth produced? How will you keep each of these pH values constant while you carry out the experiment?
7. Imagine yourself doing each experiment. What equipment and what solutions will you use in addition to those you used in Experiment 4.1?
8. How will you record your results? If you decide to use a table, you should draw it before you start your experiment and perhaps check with your teacher that it is suitable. Appendix A helps you design a table.

RESPIRATION

Every living cell needs a supply of energy to stay alive. To grow and to move it needs even more energy. The energy is supplied to cells in their food. The cells of all living organisms can release the energy in their food by a process called **respiration**. Respiration is remarkably similar in all living organisms.

Although the food of living organisms varies, by the time it is taken into a cell it is usually in the form of glucose, amino acids, fatty acids and glycerol. If the cells do not need the food immediately, they store what they can in the form of glycogen, starch or lipid.

The cell is able to change stored food into a substance from which it can release a continuous supply of energy. Glucose, one of the substances from which the cell releases energy, is a **respiratory substrate**. (Section 4.2 explains that a substrate is a substance on which an enzyme acts.) Respiration is a series of chemical reactions in which each reaction, controlled by its own enzyme, is followed quickly by the next reaction (see Figure 4.3). More than 70 different enzymes are used in respiration. That means there are more than 70 different reactions.

(a) Aerobic respiration

A glucose molecule must be broken down to release its energy. With oxygen it can be broken down completely to carbon dioxide and water which releases all its usable energy. This chemical reaction can be shown in words as

$$\text{glucose} + \text{oxygen} \rightarrow \text{carbon dioxide} + \text{water} + \text{energy}$$

and in chemical symbols as

$$C_6H_{12}O_6 + 6O_2 \rightarrow 6CO_2 + 6H_2O + \text{energy}$$

The two waste products, carbon dioxide and water, which contain no usable energy, are passed out of the cells.

Respiration in which oxygen is used to release energy is **aerobic**. Aerobic respiration occurs in animals, including humans, in plants, in bacteria and in fungi.

(b) Anaerobic respiration

Oxygen is not always present around the cells of living organisms. All cells are able, at least for a short time, to respire, and so release some energy, without oxygen. Some cells, even when oxygen is present, carry out respiration without it.

53

If we run fast, our muscles need energy very quickly. Some oxygen is stored in our muscles, and blood brings fresh supplies of oxygen all the time. But, if we run fast for long enough, there is not enough oxygen to give our muscle cells the energy they need by aerobic respiration alone. Glucose is then broken down to release energy without oxygen. Respiration without oxygen is **anaerobic**. The amount of energy released in anaerobic respiration is far less than in aerobic respiration.

In anaerobic respiration a glucose molecule is only partly broken down: it does not break down all the way to carbon dioxide and water. In the cells of humans, glucose is broken down to the waste produce **lactic acid**. This chemical reaction can be shown in words as

$$glucose \rightarrow lactic\ acid + energy$$

While some of the energy in glucose is released in anaerobic respiration, most of it goes into the lactic acid.

Lactic acid is a common waste product in anaerobic respiration: it is also formed by the action of bacteria when milk goes sour, when milk is used to form yoghurt and cheese, and when grass is converted to silage. In all these examples the chemical reaction is the same:

$$glucose \rightarrow lactic\ acid + energy$$

Another common waste product of anaerobic respiration is **ethanol** (popularly called alcohol). Yeast, a fungus, forms ethanol in anaerobic respiration. When ethanol is formed, there is also some carbon-dioxide gas given off. Again most of the energy goes into the waste product, this time ethanol. This chemical reaction can be shown in words as

$$glucose \rightarrow ethanol + carbon\ dioxide + energy$$

(c) Fermentation

Any respiration with a waste product other than carbon dioxide and water is called **fermentation**. Anaerobic respiration which breaks down glucose to lactic acid or to ethanol and carbon dioxide is fermentation. In fact all anaerobic respiration is fermentation. Aerobic respiration is also fermentation when it does not complete the process of breaking down glucose to carbon dioxide and water.

The formation of vinegar (acetic acid) from ethanol by acetic-acid bacteria (see Section 6.4) takes place only with oxygen and is therefore aerobic respiration. Because vinegar is a waste product other than carbon dioxide and water, it is also fermentation.

All waste products of anaerobic and aerobic respiration, including lactic acid, ethanol and vinegar (acetic acid), are passed out of the cell.

When cheese is made, when food goes bad and when dead bodies of plants and animals rot, it is as a result of fermentation by different bacteria and fungi. These bacteria and fungi, whether they use oxygen or not, form a number of strong-smelling waste products. Eventually aerobic respiration by other bacteria and fungi converts the waste products of these fermentations to carbon dioxide and water.

Aerobic respiration, including aerobic fermentation, takes place in a compost heap (see Section 10.5).

5.3 Usable energy

Respiration releases energy at several stages in a series of chemical reactions. Even so, about three-quarters of the energy released is in the form of heat and cannot be used by the cell for other purposes. The cell does not need all this heat. It may not need any of it. Three-quarters seems a lot of energy to lose, but most of the engines designed by humans, such as those used in cars and lorries, lose proportionately more energy than is lost in respiration.

The non-heat energy released in respiration is stored by the cell in a ready-to-use chemical substance called **adenosine triphosphate** or **ATP** for short. From one molecule of glucose, aerobic respiration makes 38 molecules of ATP but anaerobic respiration makes only two molecules of ATP. Both processes are shown in Figure 5.1.

The energy in ATP can be used by the living cells of all organisms to do work of different kinds. It is used to keep the cells alive. It is used for growth and for repair of damaged parts, which means it is used to make the thousands of different substances that cells need. In animals, including humans, and in bacteria with flagella, a great deal of the energy in ATP is used to produce movement. In birds and mammals, which are warm-blooded, the heat that is a by-product of respiration is not entirely wasted because it helps to keep the body's temperature above that of its surroundings.

5.4 Respiration in yeast

The small ovoid organism shown in Figure 2.1 has been important to humans for thousands of years. Egyptians recorded a form of beer production in 6000 BC.

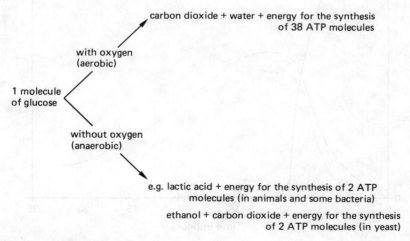

Figure 5.1   Formation of ATP

By 1200 BC they were using yesterday's dough to inoculate today's bread and to start the fermentation that makes wine. Not till 1803 did anyone suggest that microorganisms caused alcoholic (ethanolic) fermentation; in 1857 Louis Pasteur proved it.

Yeast can respire both aerobically and anaerobically. The curious fact is that, if there is plenty of glucose for yeast to feed on, it respires anaerobically even when there is plenty of oxygen. In anaerobic respiration yeast breaks down glucose, forming ethanol and carbon dioxide as its waste products. When the glucose is nearly used up, and provided that oxygen is present, yeast uses the ethanol as a respiratory substrate to produce carbon dioxide and water in aerobic respiration. This is shown by the graph in Figure 5.2.

Try following the lines of the graph in Figure 5.2. It shows the growth of a yeast colony, as measured by the cloudiness of the culture (see Section 3.3 on turbidity measurements), in a liquid medium that had air bubbled through it. The vertical scale on the right of the graph gives the cloudiness of the culture. The vertical scale on the left is in grams per cubic decimetre. At the start of the experiment the medium contained more than 40 g per dm^3 of glucose and no ethanol. After ten hours' incubation, there was virtually no glucose left in the medium and the amount of ethanol had risen steadily to 20 g per dm^3.

Yeast has behaved as you would expect (see Section 3.2); there was a lag of about five hours while yeast adjusted to its new surroundings, during which it did not grow and multiply. This was followed by about five hours of exponential growth until the glucose, the first respiratory substrate, ran out. (It is probable

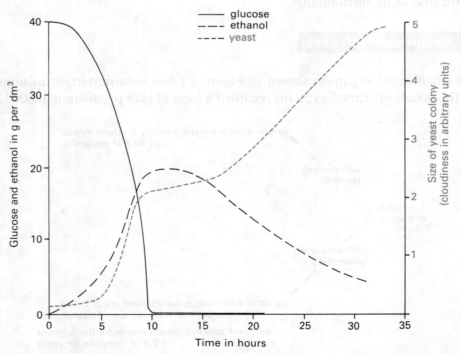

Figure 5.2 Growth of yeast in an aerobic medium

that growth stopped because there was no more glucose. It is possible that growth stopped because the ethanol poisoned the yeast cells.)

There was another lag of about five hours in the growth of yeast while it adjusted to its new surroundings in which there was no glucose but plenty of ethanol. After that the growth rate again increased. During the second lag in growth, the yeast cells made the enzymes they would use for the aerobic respiration of ethanol. After 15 hours' incubation, ethanol was used as the respiratory substrate until it too began to run out.

(a) Bread

Bread has been made for thousands of years by mixing living yeast with flour, leaving it to 'rise' (to double in volume) and then baking it. The yeast cells are usually first mixed with sugar to start their metabolic activities and in particular to get them respiring vigorously.

The respiring yeast is thoroughly mixed (kneaded) with flour, warm water and salt (which stimulates the taste cells in the tongue) to form a **dough** which is left in a warm place until its volume has doubled. Successful bread-making depends on keeping the yeast cells actively respiring: this is why warm water is added and the dough is left in a warm place to rise.

Thorough kneading spreads yeast cells throughout the dough. When the kneaded dough is left, the yeast cells respire both anaerobically and aerobically (if there is oxygen and little sugar) wherever they are within the dough. Both the anaerobic and aerobic respiration give off the waste gas carbon dioxide, which causes the dough to rise. The better the yeast cells are spread throughout the dough, the smaller and more even are the gas bubbles. Any ethanol that forms is evaporated when the bread is baked. The waste carbon dioxide is a metabolic product or **metabolite** of yeast's respiration. Figure 5.3 shows how bubbles of gas were fixed throughout the bread when the mixture was baked.

Figure 5.3 Slice of bread

Questions

Q 5.1 In Figure 5.4 a boiling tube containing a mixture of glucose and yeast in distilled water is attached to a **manometer** (a bent glass tube containing a coloured 57

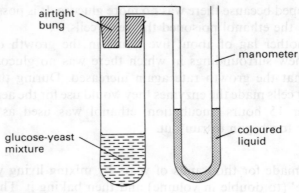

airtight bung

manometer

glucose-yeast mixture

coloured liquid

Figure 5.4   Apparatus at start

liquid) and sealed by an airtight bung. The manometer measures differences in pressure. The pressure on the surface of the liquid in the open end is the pressure of the atmosphere. The pressure on the surface of the liquid in the end attached to the boiling tube is the pressure of gas in the closed boiling tube.

(a) Half an hour after the apparatus in Figure 5.4 was set up, the liquid in the manometer moved down on the left-hand side and up on the right-hand side.
(i) What change took place in the gas in the boiling tube to make the liquid in the manometer move?
(ii) What had happened in the glucose–yeast mixture to change the gas in the boiling tube?

(b) At the start of the experiment, an apparatus was set up next to the one shown in Figure 5.4 and identical to it except for one difference: the glucose–yeast solution was first boiled and cooled. Such an apparatus, different only in **one** way from the experimental apparatus, is called a **control**. After half an hour, the liquid levels in the manometer in this control had not changed. Explain why the experiment was better when both sets of apparatus were used.

(c) In another experiment four sets, A, B, C and D, of the apparatus shown in Figure 5.4 were used. A was left at a room temperature of 18°C; B was put in a refrigerator at 0°C; C was put in an incubator at 35°C; D was put in an oven at 95°C. After half an hour the level of the manometer in

> A had fallen a little on the left-hand and risen a little on the right-hand sides
> B had risen a little on the left-hand and fallen a little on the right-hand sides
> C had fallen a lot on the left-hand and risen a lot on the right-hand sides
> D had fallen a little on the left-hand and risen a little on the right-hand sides

Suggest explanations of the results in A, B, C and D.

Q 5.2
(a) Explain what is meant by (i) respiration, (ii) aerobic respiration, (iii) anaerobic respiration and (iv) fermentation.

(b) Give **one** example of a substance formed by (i) aerobic respiration, (ii) anaerobic respiration and (iii) fermentation.

Q 5.3 Look at Figure 5.2 and answer the following questions.

(a) What was the concentration of the glucose solution at the start of the experiment?

(b) (i) What was the highest concentration of ethanol during the experiment?
    (ii) How long did it take for the ethanol to reach its highest concentration?

(c) Suggest why growth of the yeast colony slowed down 10 hours after the start of the experiment.

Practical work

Experiment 5.1 To test the hypothesis that in respiration in yeast oxygen is used and carbon dioxide is produced

Note to teachers

To make the hydrogencarbonate–indicator stock solution (in advance of the practical class) first dissolve 0.2 g of thymol-blue powder and 0.1 g of cresol-red powder in 20 cm^3 of ethanol. Add this to 0.84 g of 'Analar' sodium hydrogencarbonate dissolved in 900 cm^3 of distilled water in a 1 dm^3 volumetric flask and make up the volume to 1 dm^3 with distilled water. Cut 25 cm lengths of 5–6 mm glass delivery tubing and flame-polish the cut ends. Make two right-angle bends in each length of tubing, respectively about 5 cm and 10 cm from one end. Fit the shorter arm through a one-hole rubber bung which will fit into a 25 mm diameter boiling tube. You need two lengths of tubing for each student plus spares to allow for breakages.

On the day of the experiment, remove 3 cm^3 of the hydrogencarbonate–indicator solution per student. Dilute the volume removed with nine times its own volume of distilled water. Using an aquarium pump or filter pump, aerate this diluted solution for about fifteen minutes to get it into equilibrium with atmospheric air. It should then be red.

About 30 minutes before the practical class starts, prepare 50 cm^3 per student of the yeast–glucose mixture. To prepare 250 cm^3, add 8 g of glucose and 6 g of dried baker's yeast to 250 cm^3 of water at 45°C. Stir the mixture and leave it in a warm incubator or drying cabinet until the practical class starts. If at the end of the practical class there has not been enough aerobic respiration to change the colour of the aqueous methylene-blue solution, tell the students not to record the colours yet. It possible, arrange for the students to record them a few hours later. Otherwise record them a few hours later yourself.

Materials needed by each student

1 50 cm^3 beaker
1 10 cm^3 graduated pipette fitted with a pipette filler
1 dropper pipette
1 boiling-tube rack with 4 boiling tubes

59

2 one-hole rubber bungs with fitted bent glass delivery tubing
Bunsen burner, tripod and gauze
1 spirit marker
30 cm^3 hydrogencarbonate–indicator solution
2 cm^3 aqueous methylene-blue solution
50 cm^3 yeast-glucose mixture

Instructions to students
1. Label the boiling tubes A, B, C and D.
2. Pipette 20 cm^3 of yeast–glucose mixture into tube A.
3. Pour the rest of the yeast–glucose mixture into the beaker and boil it using the Bunsen burner, tripod and gauze while you carry out the next instruction.
4. Copy the table.

	Colour at start	Colour at end
Contents of boiling tube A		
Contents of boiling tube B		
Contents of boiling tube C		
Contents of boiling tube D		

5. Cool the boiled yeast–glucose mixture and pipette 20 cm^3 into tube B.
6. Put a few drops of methylene-blue solution into tubes A and B: the mixture should turn a definite blue. Record the colour in your table.
7. Fit the rubber bungs with bent glass delivery tubing in tubes A and B.
8. Pipette 10 cm^3 of the hydrogencarbonate–indicator solution into tubes C and D. Record the colour of the solution in tubes C and D in your table. Put the end of the bent glass delivery tubing from tube A into the hydrogencarbonate–indicator solution in tube C and the bent glass delivery tubing from tube B into the hydrogencarbonate–indicator solution in tube D.
9. Record the colours of the contents of tubes A to D at the end of the practical class (or later if your teacher tells you).

Interpretation of results
1. What effect does boiling have on the yeast cells?
2. What colour changes did you observe in tubes A and B?
3. Methylene blue loses its colour as oxygen is used up. Use this information to explain your observations in 2.
4. What colour changes did you observe in tubes C and D?
5. Hydrogencarbonate–indicator solution is red when alkaline and yellow when acid. Carbon dioxide is an acidic gas. Use this information to explain your observations in 4.
6. Explain why the boiled yeast–glucose mixture was used in these experiments.
7. Was this experiment able to tell you whether yeast had carried out anaerobic respiration? Explain your answer.

Experiment 5.2 To test the hypothesis that yeast cells produce heat during respiration

Note to teachers

The 'Instructions to students' assume that each of them will design the experiment and hand in an outline of the method and a list of practical requirements. It does not follow that the students must do the experiment individually. Because a yeast culture must be left for hours or days to produce a significant change in temperature, you may prefer the class to do collectively a single experiment which will take up less space in the laboratory. Whatever you decide, try to get the students back to the laboratory to record a temperature change themselves.

Instructions to students

1. Design an experiment to test the hypothesis that respiring yeast cells produce heat. Some clues to help you design your experiment are given below. If after reading them and thinking about them you do not know how to start, ask your teacher for help.
2. Make a list of the steps you will take during your experiment. Show it to your teacher, who will check that your method works.
3. Make a list of the apparatus and solutions you will need to do your experiment. Give it to your teacher in advance so that everything you need can be got ready.
4. When you have done your experiment, write a report telling what you did, what the results were and what these show about the hypothesis you are testing. Appendix B helps you write a report.

Clues to help you design your experiment

1. There are several ways of testing the hypothesis. You need not worry if your way is different from that of other students in your class.
2. How can you measure changes in temperature in respiring yeast cultures?
3. If all the heat produced by a respiring culture is lost to the surroundings, no rise in temperature occurs. How could you stop this heat loss?
4. Small changes in temperature are difficult to measure. Will you need a large or a small volume of yeast culture to produce a measurable change in temperature?
5. How long will you need to leave the yeast culture to produce a measurable change in temperature?
6. What must you include in the yeast culture so that the yeast can respire?
7. How can you be sure that any change in temperature is caused by the action of live yeast?
8. How often will you measure the temperature?
9. Imagine yourself doing your experiment. What equipment and what solutions will you use?
10. How will you record your results? If you decide to use a table, you

61

should draw it before you start your experiment and perhaps check with your teacher that it is suitable. Appendix A helps you design a table.

Experiment 5.3 To test the hypothesis that temperature affects the rising of a flour-yeast dough

Note to teachers

About 30 minutes before the practical class starts, prepare 20 cm^3 per student of the yeast–glucose mixture. To prepare 250 cm^3, add 8 g of glucose and 6 g of dried baker's yeast to 250 cm^3 of water at 45°C. Stir the mixture and leave it in a warm incubator or drying cabinet until the practical class starts.

The boiling tubes of risen dough will be difficult to clean unless you encourage students to rinse them out into a bucket as soon as the experiment ends.

Materials needed by each student
1 250 cm^3 beaker
1 10 cm^3 graduated pipette fitted with a pipette filler
1 boiling-tube rack with 3 boiling tubes
1 weighing bottle
1 glass rod
1 stopclock
1 thermometer
1 spatula
1 ruler graduated in mm
1 spirit marker
20 g plain flour
20 cm^3 yeast–glucose mixture

Materials needed by the class
balances to weigh 15 g (1 per 4 students)
a water bath at about 45°C
a refrigerator

Instructions to students
1. Copy the table.

Treatment of tube	Temperature in °C	Height of mixture in cm		Increase in height in cm	Percentage increase in height
		At start	At end		
Refrigerator Room temperature Water bath					

62

2. Weigh 15 g of flour and pour it into the beaker.
3. Pipette 15 cm^3 of yeast–glucose mixture into the flour and stir it all with the glass rod to produce a smooth dough.
4. Pour the same amount of dough into each of three boiling tubes. Pour the dough carefully so that it does not touch the walls of the boiling tubes. If any does stick to the walls, wipe it off. You need not transfer all the dough to the tubes.
5. Use the spirit marker to make a line on each tube at the level of the dough. Put one tube of dough in a refrigerator, put one in the water bath and leave one in the rack on the bench. Start the stopclock.
6. Ten minutes before the end of this practical class or when one of the tubes fills with dough, whichever is the sooner, mark the new level of the dough in each tube.
7. Measure the temperature of the refrigerator, the air in the room and the water in the bath and record them in your table.
8. Use a ruler to measure the heights of the marks from the bottom of each tube and record these in your table.
9. Calculate the percentage increase in the height of each mixture and enter it in your table. You calculate the percentage increase in the height by using the formula:

$$\text{percentage increase in height} = \frac{\text{increase in height in cm at end of experiment} \times 100}{\text{height in cm at beginning of experiment}}$$

Interpretation of results
1. Why was there a change in the height of the dough in each tube?
2. Explain the differences in the percentage increases in height at the three different temperatures.

FERMENTATION

6.1 Introduction

Respiration is the release of energy in living organisms. Fermentation is respiration with a waste product other than carbon dioxide and water. Fermentation is common in microorganisms. Its waste products vary greatly but always include at least one complex chemical compound such as lactic acid or ethanol (alcohol). You do not need to remember them, but the chemical formula of lactic acid is $CH_3CHOHCOOH$ and of ethanol is CH_3CH_2OH. Humans have found uses for only a few of the complex compounds that are waste products of fermentation. Ethanol (alcohol) is one you will have recognised. To produce such compounds industrially, humans make use of microorganisms. Scientists have been able to extract some of the enzymes used in fermentations from the microorganisms that make them. Some industrial fermentations use only extracted enzymes, not whole microorganisms.

This Unit is about some of the ways humans have used fermentations of microorganisms. It includes some of the earliest known examples of **biotechnology**. Technology is any deliberate use of materials or energy to meet human needs. Biotechnology is the deliberate use of biological agents to meet human needs.

6.2 Beer

A form of beer was made from barley by the ancient Egyptians in about 6000 BC. Other ancient civilisations also made beer from rice, millet, oats or other cereal grains. Today beer is made on a larger scale but the principles are the same: starch is converted to sugar and sugar is fermented by yeast to form ethanol. The raw materials for making beer in the UK are

○ malt, from barley grain, to provide food for yeast and to give flavour and colour to the beer
○ hops to give beer its bitter flavour
○ yeast for fermentation to produce ethanol
○ water

Figure 6.1 shows the main stages in the industrial brewing of beer.

(a) Malt

The cereal grain used for beer-making in the UK is usually barley. Yeast cannot use starch as food. The starch in the barley must be converted to a mixture of sugars, by enzymes in the barley itself, before yeast can use it. This process starts with **malting**. The dry barley grain is mixed with water before being spread on a   65

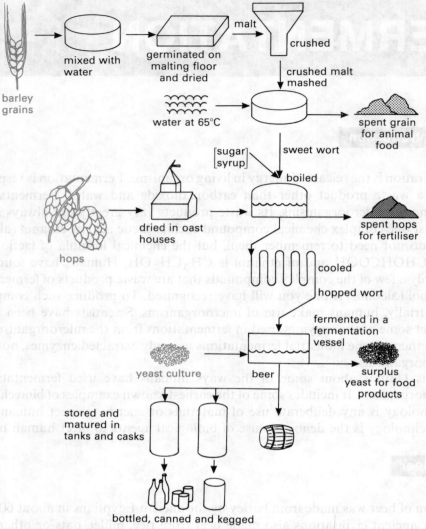

Figure 6.1   Brewing of beer

malting floor under carefully controlled conditions of temperature and moisture to make it germinate (begin to grow). The enzymes in the barley which are needed to convert starch to sugar become active during germination. After four or five days, when the enzymes have converted only some of the starch to sugar, germination is stopped by heating the grain to dry it and so stop further action of its enzymes. Drying the grain also adds flavour and colour to the end product, which is called **malt**.

The malt is now **crushed** and mixed with hot water at a temperature of about 65°C for about two hours. This is called **mashing**. A temperature of 65°C is hot but nowhere near boiling: it is not too hot to prevent the action of the enzymes. During mashing the enzymes in the crushed malt convert most of the remaining starch to sugar and the sugar dissolves in the hot water to form **sweet wort**. Mashing is shown in Figure 6.2. The sweet wort is filtered and the insoluble part

Figure 6.2   Mashing to form sweet wort

or **spent grain** is used as animal food. The brewer who makes the beer may add extra sugar, syrups or other minor ingredients before boiling the sweet wort with hops for about two hours.

(b) Hops
The **hop** is a vigorous climbing plant growing as a weed in English hedges. It is grown as a crop in the Midlands and south-east England. Hops used in brewing beer are clusters of female flowers that have been dried in **oast houses** and are now often powdered to make them easier to transport. Boiling the sweet wort with hops for about two hours sterilises it and extracts the bitter compounds from the hops. The mixture, now called **hopped wort**, is separated from the **spent hops** (which are used as fertiliser), cooled to a suitable temperature for fermentation and passed to the fermenting vessel. The hops are not essential but they make beer keep better and give it the slightly bitter taste most beer-drinkers like. Before hops were used, the drink made by fermenting barley malt was called ale.

(c) Yeast
Air or oxygen is added to the hopped wort before it is inoculated with a carefully chosen yeast culture. Each beer is made from a certain strain of yeast which gives it its flavour and strength. Brewers take care to keep their yeast free of the wild-yeast cells present in air because these would affect the flavour of the beer.

Inoculation is followed by the rapid growth of yeast and the rapid fermentation of sugars to ethanol (see Figure 5.2). The bubbles of carbon-dioxide gas given off during fermentation both give the beer a fizz and help to mix it. After about five days the hopped wort has become **beer**. Its ethanol content is between 3 and 6 per cent. The beer is not yet fit to drink: it is filtered into storage tanks or casks   67 where it matures (clears and develops more fizz and flavour). Some of the yeast that grew in the fermenting vessel is recycled to inoculate more hopped wort. The rest is used in various food products.

Genetic engineers are hard at work to improve the strains of yeast used by brewers. They even hope to engineer a strain that will act directly on the starch in the barley: this would mean that the starch need not be converted to sugar and would do away with the malting process. Ethanol poisons yeast at a concentration of about 15 per cent: if yeast could be engineered to tolerate a higher ethanol concentration, more ethanol could be produced at the same cost.

(d) Water

A good supply of fresh clean water is needed. The source of water, which the brewers call *liquor*, and the ions it contains have some effect on the flavour of the beer. The quality of the local water is one reason why brewing was concentrated in certain places such as Burton-upon-Trent (in Staffordshire) and London. Nowadays the quality of water can be changed: the process is called Burtonising.

6.3 Wine

Beer and wine are both alcoholic drinks made with the microorganism yeast. While beer is usually made from barley, wine is usually made from grapes. There are also differences in the way they are produced. There is no need in wine-making to convert starch to sugar for the yeasts to feed on: wine is made from grapes which may contain as much as 30 per cent sugar in their flesh.

Like beer, wine has been made for centuries. The grape vine is a single species with more than 5000 named varieties. The different varieties give different flavours to wines but the intensity of the flavour depends on how many grapes each vine is allowed to produce. Twelve tonnes of grapes from one hectare of vines is a suitable yield: if 24 tonnes of grapes were allowed to grow on a hectare, the wine would have only about half the flavour.

The quality of the wine is also affected by the nature of the soil. The best wines are made from vines grown in poor topsoil and well-drained subsoil: the roots must be able to grow deep and must not be drowned in water. It is said that old vines give the best wine: their roots are deepest.

The traditional way of making wine is to cut the grapes by hand (sometimes individual grapes are chosen) and to crush them by treading them barefoot in a wooden fermenting vat. Fermentation is then carried out by natural yeasts present on the skins of grapes, in the air and in the vat: no one even knows the many different varieties of these yeasts. This is different from the brewers' careful control of the yeast used in beer-making.

Wine is now drunk in such huge quantities that much of it is mass-produced in wine factories or **wineries** from grapes cut by machines. To keep the quality constant, pure cultures of yeast are used as in brewing. In Figure 6.3 you can follow the wine-making processes in a modern winery. When no more red wine runs freely out of the fermentation tank, the grape skins are transferred to a press which squeezes even more wine out of them, but this wine is of poor quality.

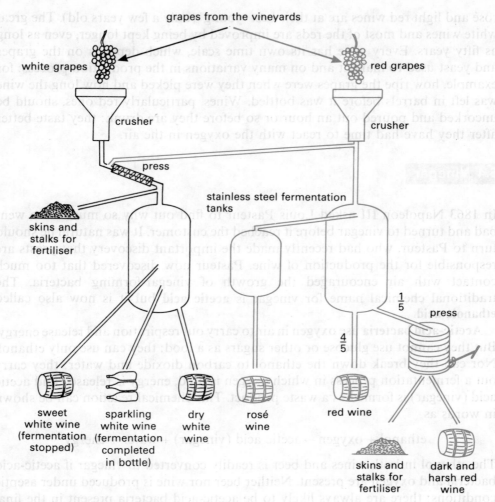

Figure 6.3 A modern winery

Left to itself, yeast normally ferments all the sugar in grapes to ethanol: this results in a **dry** wine. To make a **sweet** wine, the fermentation must be stopped earlier, by filtering off the yeast cells, by adding sulphur dioxide to stop the action of yeast, or by adding more ethanol to poison the yeast cells. The yeast cells are poisoned and fermentation stops when the ethanol content reaches about 15 per cent of the total volume. If the grapes are very sweet to start with, the ethanol content will reach 15 per cent and fermentation will stop before all the sugar has been used up: the wine will be slightly sweet. Very sweet grapes can be used to make very sweet wine by stopping the fermentation earlier.

White wine comes from both white grapes and red grapes. Most wine comes from red grapes. If the red skins are removed as soon as the grapes are crushed, the wine will be white. If the red skins are left with the crushed grapes for a matter of days, the wine will be tinted red (**rosé**). If the red skins are removed only after fermentation, the wine will be **red**. Complete fermentation takes about 14 days.

Some wines are ready to drink almost as soon as they are made. Most white, rosé and light red wines are at their best 'young' (only a few years old). The great white wines and most of the reds are improved by being kept longer, even as long as fifty years. Every wine has its own time scale, which depends on the grapes and yeast used to make it and on many variations in the production process: for example, how ripe the grapes were when they were picked and how long the wine was left in barrels before it was bottled. Wines, particularly red ones, should be uncorked and poured out an hour or so before they are drunk: they taste better after they have had time to react with the oxygen in the air.

6.4 Vinegar

In 1863 Napoleon III asked Louis Pasteur to find out why so much wine went bad and turned to vinegar before it reached the customer. It was natural he should turn to Pasteur, who had recently made the important discovery that yeasts are responsible for the production of wine. Pasteur now discovered that too much contact with air encouraged the growth of vinegar-forming bacteria. The traditional chemical name for vinegar is **acetic acid** but it is now also called **ethanoic acid**.

Acetic-acid bacteria use oxygen in air to carry out respiration and release energy. But they cannot use glucose or other sugars as a food: they can use only ethanol. Nor can they break down the ethanol to carbon dioxide and water: they carry out a fermentation process in which oxygen is used, energy is released and acetic acid (vinegar) is formed as a waste product. This chemical reaction can be shown in words as

$$\text{ethanol} + \text{oxygen} \rightarrow \text{acetic acid (vinegar)} + \text{water} + \text{energy}$$

The ethanol in both wines and beer is readily converted to vinegar if acetic-acid bacteria and oxygen are present. Neither beer nor wine is produced under aseptic conditions: there are always likely to be acetic-acid bacteria present in the final product. This is why beer and wines will not keep for long once they are exposed to air.

6.5 Yoghurt

Experiment 6.2 tells you how to make yoghurt from milk. Making yoghurt is another example of biotechnology that humans have used for centuries. What you will do on a small scale is what the yoghurt manufacturer does on a large scale. An outline of the process is shown in Figure 6.4.

You will start with cows' milk that is **homogenised**: even and smooth in texture, with the fat mixed well into it. Manufacturers have first to **homogenise** cows' milk. They add milk solids and, if they wish to make sweet yoghurt, they also add sugar syrup. The mixture is partially sterilised by keeping it at about 90°C for 15 to 30 minutes. Complete sterilisation would be expensive and would spoil the yoghurt's flavour and texture.

70   When it has cooled to 44°C, it is pumped into vats for fermentation. It is

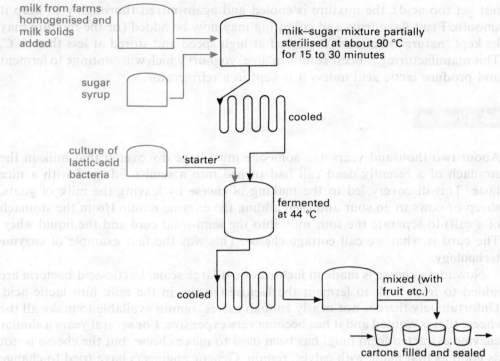

milk from farms homogenised and milk solids added

sugar syrup

milk–sugar mixture partially sterilised at about 90 °C for 15 to 30 minutes

cooled

culture of lactic-acid bacteria

'starter'

fermented at 44 °C

cooled

mixed (with fruit etc.)

cartons filled and sealed

Figure 6.4   Manufacture of yoghurt

inoculated with a 'starter', a culture of two different species of lactic-acid bacteria (*Lactobacillus bulgaricus* and *Streptococcus thermophilus*), both of which ferment some of the sugar (lactose) in milk to lactic acid. This chemical reaction can be shown in words as

$$\text{lactose sugar} + \text{water} \rightarrow \text{lactic acid} + \text{energy}$$

As the mixture's pH falls (as it becomes more acid), it thickens. Other complex fermentation products of the two species of bacteria give yoghurt its flavour. The two species of bacteria are shown in Figure 6.5: a bacillus is rod-shaped and a coccus is spherical.

After about 5 hours, when the pH has reached the right level (the mixture must

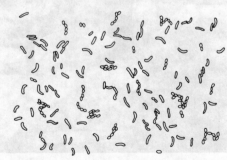

Figure 6.5   *Lactobacillus bulgaricus* and *Streptococcus thermophilus*

not get too acid), the mixture is cooled and again mixed thoroughly to keep it smooth. Fruit flavouring and colouring may now be added (or the mixture may be kept 'natural') before it is packed at high speed and stored at less than 5°C. This manufacturing process results in 'live' yoghurt which will continue to ferment and produce lactic acid unless it is kept in a refrigerator.

6.6 Cheese

About two thousand years ago someone must have discovered that milk in the stomach of a recently dead calf had turned into a semi-solid curd with a nice taste. This discovery led to the making of **cheese** by leaving the milk of goats, sheep or cows to go sour and then adding the enzyme rennin (from the stomach of a calf) to separate the sour milk into the semi-solid **curd** and the liquid **whey**. The curd is what we call **cottage cheese**. This was the first example of **enzyme technology**.

Nowadays cheese is made in factories on a large scale. Lactic-acid bacteria are added to fresh milk to ferment the lactose (sugar) in the milk into lactic acid. Unfortunately there is not nearly enough calves' rennin available to make all the cheese that is wanted and it has become very expensive. For several years a similar enzyme extracted from fungi has been used to make cheese, but the cheese is not as good as that made with calves' rennin. Genetic engineers have tried to change bacteria so that they will make calves' rennin. Other researchers have found a way of immobilising calves' rennin on porous chicken bones so that the curd and whey can be drained off after they have formed, leaving the expensive immobilised

72   Figure 6.6   Different cheeses

rennin to be used again and again (see Section 4.5). The curd can be eaten without further processing as cottage cheese.

Most cheeses are further ripened or **matured** by the addition of other microorganisms. Figure 6.6 shows a few of the thousands of different cheeses that are produced throughout the world. Different flavours depend partly on which milk is used; in Europe it is usually from cows, sheep or goats, but milk from buffaloes, yaks and camels is used where these animals are common. Stronger flavours are provided by the many different microorganisms used for further fermentations of the milk proteins and fats: for example, the microorganisms that we see as blue-green moulds grow through the curd to make the blue-vein cheeses. The manufacture of a blue-vein cheese is shown in Figure 6.8.

The cheeses in Figure 6.6 have had the curd put into containers and pressed to remove excess moisture before being inoculated with the microorganisms. Sometimes the microorganisms grow only over the surface while the enzymes they secrete mature the inside of the cheese. Further flavours may be added in other ways: for example, the cheeses may be smoked or covered in herbs. The Swiss cheeses with holes in Figure 6.6 have been inoculated with bacteria that give off large amounts of carbon-dioxide gas during fermentation.

6.7 Silage

We use fermentation to make food for animals as well as for ourselves. **Silage** is made from living green plant material, mainly grass cut before the flowers form. The starches and sugars in the plant are fermented in the absence of air by bacteria, in particular by lactic-acid bacteria, to form lactic acid which preserves the grass as food. In winter when grass has stopped growing, silage is given to cattle and other animals that chew the cud (chew food that has already been in their stomachs) giving bacteria plenty of time to digest cellulose.

Silage also contains valuable food compounds resulting from the action of the microorganisms' enzymes, not only on starches and sugars, but on other substances such as cellulose.

Figure 6.7 shows two tower silos in which grass can be fermented in the absence of air. Fresh-cut grass is blown up the side shute and into the silo at the top. There are always lactic-acid-forming bacteria on grass to start the fermentation. A silo gives them more or less anaerobic conditions because it lets in no air (and hence no oxygen) at the sides and because the weight of the grass above compresses the grass below. It is important that other acid-forming bacteria are discouraged: butyric-acid-forming bacteria, for example, give rise to foul-smelling products and spoil the silage.

The skill in managing a silo is to ensure that air is kept out by correct packing, that the temperature is suitable for the right bacteria and not for the wrong ones, and that the pH is kept acid. Some farmers add sugar syrup (molasses) to start fermentation by the right lactic-acid-forming bacteria. Some add sulphuric acid to make conditions suitable for the right lactic-acid-forming bacteria. Good silage smells sweet and is green and juicy: it is taken out from the bottom of the silo to be fed to the animals in winter when they are usually kept under cover.

Figure 6.7   Soviet tower silos

In the UK, tower silos are out of favour. Silage can be made more cheaply and at least as well by packing the cut grass into pits, by compressing it to get rid of the air, by driving a tractor over it, and by covering it with polythene sheets weighted down with old tyres to keep the air and rain-water out. Another method is to seal tight bundles of cut grass in polythene bags. These bags weigh up to half a tonne and are handled mechanically.

Questions

Q 6.1. Look at Figure 6.1, which shows stages in the brewing of beer, and answer the following questions.
(a) Why is the barley grain germinated before it is crushed?
(b) What happens during mashing when the crushed malt and water are kept at 65°C for about two hours?
(c) Why does the brewer boil sweet wort with dried hops?
(d) What happens in the fermenting vessel when a culture of yeast is added to hopped wort?
(e) (i) Name the three by-products of the brewing of beer.
   (ii) What use is made of each of these three by-products?

Q 6.2. Look at Figure 6.3, which shows stages in the production of wine, and answer the following questions.
(a) How is white wine obtained from red grapes?
(b) How is rosé wine obtained from red grapes?
(c) What happens in the fermentation tanks?
(d) Why does fermentation stop after about two weeks in the fermentation tanks?

Q 6.3. What is the role of yeast in (a) the brewing of beer and (b) the making of wine?

Q 6.4. How is the formation of vinegar prevented in beer and wine?

Q 6.5. Look at Figure 6.4, which shows stages in the manufacture of yoghurt, and answer the following questions.
(a) Why is the milk-sugar mixture kept at 90°C for 15 to 30 minutes?
(b) Why is a 'starter' added to the cooled milk–sugar mixture?
(c) Why is the temperature in the fermenting vessel kept at 44°C?

Q 6.6. Figure 6.8 shows stages in the manufacture of a blue-vein cheese.
(a) Why are (i) lactic-acid bacteria and (ii) rennin added to sheep's milk?
(b) Why is a culture of *Penicillium roqueforti* added to the curd?
(c) What happens during the months when the cheese is maturing?

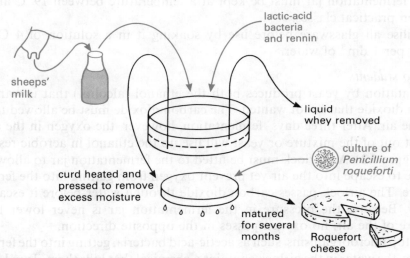

Figure 6.8   Manufacture of a blue-vein cheese

Q 6.7.
(a) What are the right conditions for the production of good sweet-smelling silage?
(b) Why must (i) air be excluded as far as possible from the cut grass and (ii) the temperature not be allowed to get very high in a silo?
(c) Why does the temperature rise in a silo?

Q 6.8. Beer, wine, vinegar, yoghurt, cheese and silage are all made from fermentations. Which of these fermentations is aerobic respiration? Explain your answer.

Experiment 6.1 To show fermentation by yeast with the use of an airlock
Note to teachers
The materials needed to carry out this experiment can be obtained from homebrew suppliers or from larger branches of High Street pharmacies and drug stores.

Since the fermentation jars are cumbersome and take up valuable laboratory space, it is suggested that a single experiment will suffice for the whole class. Older students may already make their own beer or wine at home. If so, one or two of them may like to bring their equipment to show the rest of the class how it is used.

The experiment lasts two weeks. After three days the students should, if it is convenient, measure the specific gravity and insert the airlock themselves. If it is not convenient, you should insert the airlock after three days yourself. The students can then measure the specific gravity of the fermenting mixture during the next practical class and again during the practical class one week after that. Allow about 15 minutes for the students to measure the specific gravity. Students can carry out this experiment at home.

The fermentation jar must be kept at a temperature between 19°C and 24°C between practical classes.

Sterilise all glassware before use by soaking it in a solution of 4 Campden tablets per 1 dm³ of water.

Note to students
Fermentation by yeast produces both the ethanol (alcohol) that is wanted and carbon dioxide that is not wanted. The carbon dioxide must be allowed to escape into the air. After three days' fermentation, however, the oxygen in the air must be kept out of the mixture or yeast will use up the ethanol in aerobic respiration (see Figure 5.2). An airlock must be fitted to the fermentation jar to allow carbon dioxide to escape into the air yet prevent oxygen from getting into the fermenting mixture. The airlock passes carbon dioxide through water before it escapes into the air. Because the pressure in the fermentation jar is never lower than the pressure of the air, no oxygen passes in the opposite direction.

To stop microorganisms, such as acetic-acid bacteria, getting into the fermenting mixture, the water in the airlock contains a chemical that kills them. Two U-shaped airlocks are shown in use in Figure 6.9. Any kind of airlock can be used.

Figure 6.9 Two demijohns of fermenting mixture fitted with airlocks

Materials needed to set up the fermenting mixture
1 4.5 dm^3 glass fermentation jar (demijohn) with fitted airlock
1 1 dm^3 beaker
1 250 cm^3 beaker
1 glass funnel
1 glass rod
1 thermometer
non-absorbent cotton wool to make a loose plug for the fermentation jar
1 kg sucrose
4 g citric acid
4 g yeast nutrient
2 Campden tablets
4.5 dm^3 boiled water cooled to 24°C
1 standard sachet of brewer's yeast

Materials needed to measure the specific gravity of the fermenting mixture
1 250 cm^3 measuring cylinder
1 hydrometer to measure specific gravities of 1.000 to 1.150
1 siphon or length of clean tubing to act as a siphon

Instructions to students
1. Put the citric acid and sucrose in the 1 dm^3 beaker. Add about 500 cm^3 of the boiled water and stir the mixture to dissolve the citric acid and sucrose. Add more water if it is needed.
2. Remove the airlock from the fermentation jar and use the filter funnel to pour the solution of sucrose and citric acid into the jar.
3. Mix the yeast and yeast nutrient with a little of the boiled water. Pour this mixture into the fermentation jar. Add the rest of the boiled water to the fermentation jar.
4. Use the siphon to fill the measuring cylinder with the mixture. Put the hydrometer, heavy end downwards, into the mixture in the measuring cylinder and rotate it quickly to get rid of any air bubbles clinging to it. When the hydrometer is floating perfectly still and is not touching the side of the cylinder, read off the specific gravity of the mixture. Record the specific gravity and the date in a table of your own design. Appendix A helps you design a table. Throw away the mixture in the measuring cylinder.
5. Add a Campden tablet to the mixture in the fermentation jar and plug the jar with cotton wool. Leave the fermentation jar at a temperature between 19°C and 24°C.
6. After three days, if you have access to the laboratory, remove the cotton-wool plug. Take a sample of the fermenting mixture and read off its specific gravity by repeating instruction 4. Half fill the airlock with water containing one-eighth of a Campden tablet and insert the airlock in the jar.
7. After another four days remove the airlock. Take a sample of the fermenting mixture and read off its specific gravity by repeating instruction 4. Replace the airlock.
8. After another week remove the airlock. Take a final sample of the fermenting mixture and read off its specific gravity by repeating instruction 4.

77

Interpretation of results

1. Calculate the change in specific gravity between your first and last reading. You can use this figure to calculate the percentage of ethanol (by volume) of the fermented liquid by using the formula

$$\frac{\text{initial specific gravity} - \text{final specific gravity}}{0.00736} = \begin{array}{l}\text{percentage}\\ \text{ethanol}\\ \text{by volume}\end{array}$$

Some hydrometers do not have a decimal point on their scales: e.g. the specific gravity of water reads 1000 instead of 1.000. If your hydrometer readings have values between 1000 and 1150 divide by 7.36 (instead of 0.00736) in the above equation.

2. Why did the airlock have a solution of Campden tablet in it?
3. It is not possible for yeast to produce a solution with an ethanol concentration of more than 15 per cent (by volume). Explain why.

Experiment 6.2 To make yoghurt
Note to teachers

Students can safely do this experiment at home. If it is done in a laboratory, they must be given the opportunity to examine the finished yoghurt the following day. Students should not be allowed to eat yoghurt which has been made in a laboratory.

Note to students

If you carry out this experiment at home, you will need a vacuum flask or a wide-necked container, such as a plastic cup, in which to put the milk. The culture should be kept in a vacuum flask or in a warm place such as an airing cupboard with a hot-water storage cylinder. You may be able to get hold of an electric yoghurt-maker which has its own wide-necked containers and can be kept at a constant warm temperature.

If you like to eat yoghurt, you can adapt this method to make larger quantities. Yoghurt can be made from skimmed or full-fat milk from cows, sheep or goats. If you prefer thick yoghurt, add some skimmed-milk powder to the milk before adding the 'starter'.

You can use some of each batch of yoghurt you make as a 'starter' for the next batch. After you have done this four or five times, however, the proportions of the two species of bacteria which make the yoghurt will have changed and you will need to start with a fresh supply. Once you have made the yoghurt, you should keep it in a refrigerator until you eat it.

Materials needed by each student
1 250 cm^3 beaker
1 spirit marker
1 plastic teaspoon
plastic cling film to cover the beaker
150 cm^3 freshly opened UHT milk
10 cm^3 unpasteurised natural yoghurt

Materials needed by the class
an incubator at 40°C

Instructions to students
1. Write the date and your initials on the beaker.
2. About half fill the beaker with UHT milk.
3. Stir a teaspoonful of yoghurt into the beaker of milk. This is the 'starter'.
4. Cover the beaker with cling film and leave it on the shelf of the incubator at 40°C.
5 Examine the contents of the beaker the next day and notice its smell and consistency. Its consistency is how thick or thin it is.

Interpretation of results
1. Why were you told to use UHT milk?
2. Why was the mixture of milk and 'starter' yoghurt left at 40°C?
3. How had the milk mixture changed after one day?
4. Humans have made yoghurt for centuries. The practice started in hot countries where milk quickly sours. Suggest why milk turns sour more quickly in hot countries and why turning milk into yoghurt helps to preserve it.

Experiment 6.3 To test the hypothesis that yoghurt is made as a result of the action of microorganisms on milk
Note to teachers
The 'Instructions to students' assume that each of them will design the experiment and hand in an outline of the method and a list of practical requirements. Students must do Experiments 3.1 and 6.2 before they do this one.

Instructions to students
1. Design an experiment to test the hypothesis that milk is turned into yoghurt by the action of microorganisms in the 'starter' yoghurt. Some clues to help you design your experiment are given below. If after reading them and thinking about them you do not know how to start, ask your teacher for help.
2. Make a list of the steps you will take during your experiment. Show it to your teacher, who will check that your method works.
3. Make a list of the apparatus and solutions you will need to do your experiment. Give it to your teacher in advance so that everything you need can be got ready.
4 When you have done your experiment, write a report telling what you did, what the results were and what these show about the hypothesis you are testing. Appendix B helps you write a report.

Clues to help you design your experiment
1. There are several ways of testing the hypothesis. You need not worry if your way is different from that of other students in your class.
2. Remind yourself of how you made yoghurt by reading your report of Experiment 6.2. Remind yourself too of how you showed the activity of milk microorganisms by reading your report of Experiment 3.1.

3. What volume of milk will you use? How can you ensure there are no bacteria already in it?
4. What will you need to add to a container of milk and 'starter' yoghurt to show that it contains active microorganisms?
5. Will showing that milk contains active microorganisms be enough to 'prove' that they are responsible for making the yoghurt? If not, can you think of a suitable control experiment? For example, can you destroy the micro-organisms in some of the milk?
6. For how long will you leave the mixture of milk and 'starter' yoghurt before recording the results? Under what conditions will they need to be left?
7. Imagine yourself doing your experiment. What equipment and what solutions will you use?
8. How will you record your results? If you decide to use a table, you should draw it before you start your experiment and perhaps check with your teacher that it is suitable. Appendix A helps you design a table.

NUTRITION IN MICROORGANISMS

7.1 Nutrition

All living organisms need a supply of food. Getting food, taking in food and making food (as plants do) are called **nutrition**. Organisms need food because it contains

○ energy which they can release in respiration and use in the form of ATP
○ raw materials from which they build other substances for growth and repair (building these other substances is called **biosynthesis**)

Chemical compounds from which organisms can obtain energy are nearly always **organic**. Until about 150 years ago, organic compounds could be made only by living organisms (which is why they are called organic); nowadays humans make them in their laboratories and factories. Organic compounds always contain carbon, usually contain hydrogen, often contain oxygen, and sometimes contain nitrogen, sulphur, phosphorus and other elements. Starch, sugars, fats, proteins, ethanol, acetic acid and lactic acid, which have been mentioned in earlier Units, are all examples of organic compounds. Very simple carbon compounds, such as carbon dioxide and carbonates, are included among **inorganic** compounds.

There are two main methods by which living organisms get their food:

○ plants and a few bacteria (see Sections 8.2 and 10.4) make it themselves from inorganic substances, using an outside source of energy (**autotrophic feeding**)
○ animals, fungi and most bacteria absorb it in the form of simple organic compounds (**heterotrophic feeding**)

To get their food in the form of simple organic compounds, which have small molecules and are soluble (will dissolve in water), most animals, fungi and bacteria must digest it (see Section 4.5). Most animals digest their food inside their bodies while fungi and bacteria digest their food outside their bodies (**externally**). In all these organisms, enzymes are used to carry out digestion. Section 4.5 describes the use that humans make in biological washing powders of enzymes secreted by bacteria to digest proteins.

7.2 Saprophytes

Bacteria and fungi which feed on non-living organic material and digest it outside their bodies are **saprophytes**. Saprophytes feed by absorbing simple dissolved substances through their body surfaces. If their food is insoluble, they secrete 81 enzymes on to it to digest it and convert it to a simple soluble form. Figure 4.6 shows the principles involved in saprophytic nutrition: secretion of enzymes on to insoluble food, external digestion to simpler soluble compounds and the absorption of soluble food.

(a) Diffusion

Diffusion is a physical process taking place around us all the time. When you cut open an onion, you soon smell it. People standing further away smell it a bit later than you and smell it less strongly. People still further away will not smell it at all.

When you cut open the onion, the molecules of the gas that forms the onion smell escape into the air and move off in all directions. As they escape from the onion, they get further apart, which is why people further away smell the onion less strongly. The onion molecules reach people even further away, but by this time there are so few of them that the sense cells of the nose cannot detect them.

This is diffusion. The molecules spread out from regions where there are lots of them (where they are **concentrated**) to regions where there are few of them (where they are **dilute**). This goes on until they are about equally spread out: even though they go on moving about after that, they are all moving about at more or less the same speed in all directions and remain about equally spread out.

This spreading out, or diffusion, goes on easily when one gas is spreading through another. It also goes on easily when a substance is spreading through a liquid, provided that the spreading substance is a gas or is dissolved.

When saprophytes feed, the large enzyme molecules in their cytoplasm are secreted on to the food outside: they pass out through the cell-surface membrane and the cell wall. The small molecules of soluble food are in higher concentration outside the cytoplasm: they therefore diffuse into the cytoplasm through the cell wall and cell-surface membrane. Only gases and soluble substances in solution can diffuse into and out of living cells.

(b) Food sources

Several of the previous Units have described different sources of food of microorganisms:

○ *Rhizopus* fed on starch in starch agar
○ yeast fed on sugar in sugar solution
○ lactic-acid bacteria fed on lactose sugar in milk
○ acetic-acid bacteria fed on ethanol in beer and wine

Many microorganisms are **specific** in the organic food they need: for example, acetic-acid bacteria will feed only on ethanol. This is the result of the specific nature of enzymes: that they will work only on one type of substrate. Many microorganisms make only a few digestive enzymes for secretion on to their food: the food they can use is limited to those on which their enzymes act.

Microorganisms also vary in the proportions of the food they need in an organic form. Small plants do not need any organic food at all. Many bacteria and fungi need only one or two compounds in an organic form: they can make all the substances for growth and repair with the addition only of inorganic ions. At the other extreme, some bacteria and fungi need all their food in an organic form.

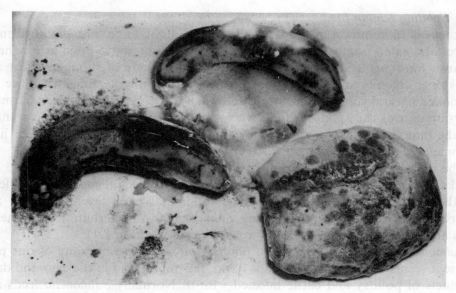

Figure 7.1 Saprophytes causing decay

7.3 Decay

When saprophytic microorganisms digest their food externally to form substances that they can absorb, they **decay**, **rot** or **decompose** it. Figure 7.1 shows some of the microorganisms that landed on two bananas and a large bread roll. The small round patches on the bread could be colonies of the fungus *Penicillium* which have grown from single spores that landed on the bread. The white fluffy patches on the banana, also growing out from the banana towards the other food, are the hyphae of other fungi. The smaller patches are likely to be colonies of bacteria and yeasts.

Rhizopus is likely to be on the bread because it feeds on starch. Yeasts are likely to be on the bananas because they feed on sugars. Staining techniques and a microscope are needed to identify all the different saprophytes that are feeding on, and causing the decay of, these foods.

At one time all organic substances were either compounds that living organisms had made or the remains of organisms that had died. All could be decayed by saprophytes for whom they were a source of food. Anything that will decay because of the action of saprophytes is said to be **biodegradable**. Now many organic compounds are made in laboratories and factories. Some, such as certain plastics, cannot be used as a source of food by any known saprophyte. They do not therefore decay: they are **non-biodegradable**.

All litter pollutes our surroundings. But when things made of paper or cardboard are left in the countryside, they do decay, usually within months. Things made of plastic, such as bags and bottles, will last until wind and rain have eventually battered them to bits. That can take years and years. Genetic engineering may one day provide a solution: microorganisms engineered to feed on plastics may be released.

83

(a) Spoiling human food

Rhizopus is so common in the air that we have all at some time breathed in its spores and eaten small colonies of it. This is unlikely to have done us any harm because *Rhizopus* feeds on starch and we are not made of starch. We eat and drink live colonies of harmless bacteria and fungi all the time when we eat uncooked natural yoghurt and cheese and when we drink milk. But even harmless microorganisms can spoil our food when they feed on it: no one would want to eat the mouldy bananas and bread in Figure 7.1 or drink fruit juices that have become acid and vinegary.

The microorganisms that harm us when they get on to our food (or into our drinks) are few but important. One of the commonest is *Salmonella*, which feeds on the lining of our intestines, causing abdominal pain, fever, vomiting and diarrhoea. We call this food poisoning. *Salmonella* is a bacterium which, when it lives and grows in our intestines, is a **parasite** (see Section 7.4).

Salmonella is found in the gut of poultry, of cattle and of many of our pets, where it seems to do them no serious harm. It is found also in the eggs and dairy products of poultry and cattle infected with it. Normally *Salmonella* feeds and grows in living organisms as a parasite but it can feed as a saprophyte on cooked and uncooked meat. In the UK about 40 people a year, mainly the elderly, die as a result of food poisoning caused by *Salmonella* infections. This is a tiny proportion of all the people who suffer from it.

Figure 7.2 shows an increase in *Salmonella* food poisoning since the early 1940s. But it shows only the cases that have been recorded by identification of the bacterium in Public Health Laboratories. Most people do not go to the doctor with *Salmonella* food poisoning. Nor do doctors often send the Public Health Laboratories samples of faeces of patients who do go to them with *Salmonella* food poisoning. There are therefore many more cases of *Salmonella* food poisoning than are shown in Figure 7.2: probably more than a million year.

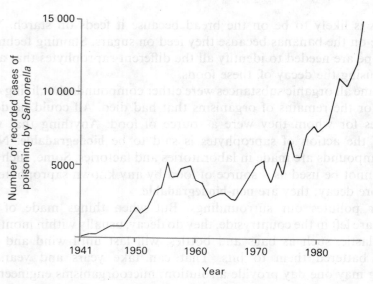

84 Figure 7.2 Recorded cases of *Salmonella* food poisoning

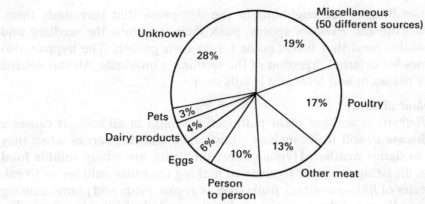

Figure 7.3 Sources of 500 *Salmonella* infections

Why has *Salmonella* food poisoning increased since the 1940s when refrigeration and the hygienic handling of food have also increased? One reason is intensive farming: keeping animals close together increases infection. Most of the frozen poultry sold in the shops was reared by intensive farming. According to the Director of the Central Public Health Laboratory, nearly all frozen poultry is infected with *Salmonella*, which is why frozen poultry must be thoroughly cooked before it is eaten.

Intensive farming is not the only source of infection: the pie chart in Figure 7.3 shows the source of infection, where it could be traced, in 500 cases of *Salmonella* food poisoning in the United States.

7.4 Parasites

Most microorganisms are saprophytes, getting their food from dead organic material which they usually need to digest externally before they can absorb it. But a small and important number of microorganisms are **parasites**. A parasite gets its food from another **living** organism which is larger than itself, which it harms in some way, but which it does not usually kill and rarely kills quickly.

Salmonella is a parasite that feeds on and harms living organisms such as ourselves. It is usual to call microorganisms that cause diseases **pathogens**. *Salmonella* is both a human pathogen and a human parasite. It is a parasite but not a pathogen of other animals such as poultry and cattle because it lives inside them, feeding off them, but does not give them food poisoning or any other disease. There are also plant pathogens, microorganisms that cause diseases in plants. Small numbers of pathogens are everywhere and cause no harm. It is estimated that it takes about 100 million *Salmonella* bacteria to cause an attack of food poisoning in a person.

(a) Pythium
The fungus *Pythium* is a plant pathogen common in all soils. It feeds on seedlings causing a disease called **damping-off**. If seeds are planted close to one another and air does not circulate freely between them, *Pythium spores* can swim from   85

one infected seedling to its neighbour in the dampness that surrounds them. Hyphae grow from the *Pythium* spores, push their way into the seedling and absorb the soluble food that it has made for its own growth. The hyphae also secrete enzymes for external digestion of the seedling's own cells. All this softens the seedling's tissues at soil level and it falls over.

(b) Other plant diseases

The fungus *Botrytis* is another plant pathogen common in all soils. It causes a **grey-mould disease** in soft fruits such as strawberries and raspberries when they are ripening in damp weather. Hyphae infect the fruits, absorbing soluble food such as sugar, digesting the inner tissues and making the fruits unfit for us to eat.

Certain species of *Rhizopus* attack fruits such as apples, pears and plums, causing a brown **soft-rot disease** of the inner tissues. Spores reach the fruits in rain splashes and grow hyphae on them. Hyphae push through the surface cells of a fruit and secrete enzymes that digest its inner flesh: they absorb both this digested food and simple soluble food stored in the fruit. The result is the brown rot called soft rot.

Plant diseases caused by fungi can ruin crops of fruit and vegetables causing great hardship to communities that depend on them as their main food. A fungal disease of potatoes in Ireland in the late 1840s caused widespread starvation: about one million people died and one and a half million emigrated to escape starvation.

Even though there are now **fungicides** (compounds that kill fungi) and sprays that prevent the growth of fungal spores, large-scale crop failures are still possible. Suitable weather conditions (such as warmth and moisture) at exactly the right time for the growth of these pathogens can make it impossible to prevent the spread of plant diseases.

7.5 Preserving human food

Preventing food from being spoiled by microorganisms is called preserving it. Today there are many ways of preserving food. Some ways, such as **refrigeration** and **pickling**, make conditions unsuitable for the growth of microorganisms. Other ways, such as **irradiation**, kill the microorganisms in food. Yet other ways, such as **canning**, protect from re-infection food in which the microorganisms have been killed. The examples that follow are only three of many different methods of food preservation.

(a) Storing food at low temperatures

Enzymes work very slowly at low temperatures, and saprophytes must use enzymes to digest their food and make the substances they need in order to grow and reproduce. Refrigerators at a temperature between 0°C and 5°C will slow down, but will not stop, decay by saprophytes. Deep-freeze cabinets at a temperature a little below − 15°C will stop decay by saprophytes but will not kill them.

Food that is thawed must not be refrozen: during the time it takes to thaw, contaminating microorganisms will have had a chance to grow and multiply. Remember that numbers make all the difference: it takes about 100 million

Salmonella bacteria to cause an attack of food poisoning. Uncooked meat that has been frozen (let alone refrozen) should always be well cooked to kill *Salmonella* bacteria that may be in it.

(b) Pickling in vinegar

Not many enzymes can work in acid conditions. Vinegar is a dilute solution of acetic acid. Soaking food in vinegar makes the pH unsuitable for those enzymes with which microorganisms carry out external digestion (because they need a near-neutral or alkaline pH in which to work). Not many foods are treated in this way because such acid foods, like pickled onions and chutney, are eaten only in small quantities by humans.

(c) Food irradiation

Large doses of **ionising radiations** break down the molecules in food and release ions. Ionising radiations kill the microorganisms in food and increase the time that it will keep without spoiling. What no one knows for sure is whether the chemical changes produced in food by ionising radiations are harmful to humans. It is likely it causes some damage, such as destruction of vitamins. Treating human food with ionising radiations was tried early this century and then abandoned. Some animal food is treated with ionising radiations.

7.6 Industrial applications of biosynthesis in microorganisms

Biosynthesis is the use of raw materials to build up new substances within the cells of living organisms. It goes on in all organisms when they grow and when they repair themselves. The raw materials are provided by their food, which must pass into the cells in a soluble small-molecule form. From these small molecules the cells build up (synthesise) the larger more complex molecules. This process needs energy in the form of ATP which is released by respiration in the cell.

In Unit 6 several human uses of fermentation processes are described. These are not biosyntheses because they break down substances with the release of energy. Biosyntheses build up substances using energy.

The fermentations described in Unit 6 take place in fermentation vessels, called **fermenters**, in which food and microorganisms have been put. During fermentation the conditions in the vessel continually change: the food is used up, oxygen may be used up, the microorganisms grow and multiply and their waste products increase. Eventually fermentation stops, usually because the food runs out or the microorganisms are poisoned by their own waste products. Such a culture of microorganisms is described as a **batch culture**. When you grow *Rhizopus* on agar you are growing a batch culture because you allow conditions in the medium to change as growth occurs.

Although batch culture is suitable for many industrial processes, it is sometimes better to grow microorganisms under constant (unchanging) controlled conditions. This process is called **continuous culture**. Figure 7.4 shows a fermenter which can be used for continuous culture. Taps on the fermenter allow
○ removal of the culture (the mixture of medium and microorganisms)
○ the addition of fresh sterile medium

87

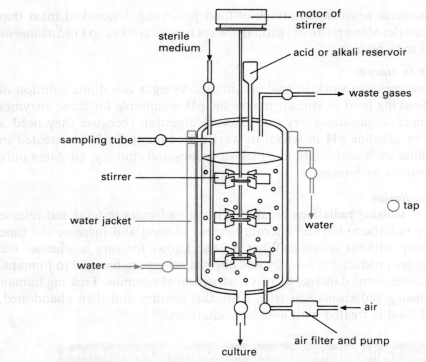

Figure 7.4   An industrial fermenter

○ the entry of air (or oxygen)
○ temperature control by an outer water jacket
○ pH control from an acid or alkali reservoir
○ continual sampling of the culture for testing

Conditions are kept the same throughout the culture by continual stirring.

Until something goes wrong, growth and multiplication of the microorganisms can be kept going because their food need not run out, their waste products are continually removed, and the temperature and pH of the culture can be controlled.

Our intestines work on a continuous-culture process. Many microorganisms grow there, some useful, some harmless: as we eat we give all these microorganisms a continual fresh food supply. This enables them to grow and multiply; part of the culture is continually removed in faeces.

(a) The production of antibiotics

Some substances produced by microorganisms prevent the growth of other microorganisms. They are called **antibiotics**. Most antibiotics are produced by fungi and prevent the growth of bacteria. The most well known antibiotic, the first to be produced and the one that is most widely used, is **penicillin**.

Penicillin is produced and secreted by the fungus *Penicillium* in the later stages of its growth. No one is sure why *Penicillium* produces penicillin. Penicillin does stop bacteria growing near *Penicillium* and so helps it to keep nearby food for itself. But this would be of most use early in the growth of *Penicillium*; most penicillin is produced when growth has almost stopped.

88   The industry world-wide produces over 100 million kilograms of penicillin each

year. Industry has had more than forty years to improve the techniques of penicillin production. The process is now automated and computerised to provide optimum conditions for the most economic production of good-quality penicillin.

Production takes place in huge fermenters, like the one shown in Figure 7.5, in which all stages can be controlled. The medium contains a mixture of glucose and lactose sugars: *Penicillium* grows better on glucose but produces more penicillin on lactose. Air is pumped at pressure into the fermenter so that oxygen is well mixed with the medium. A cooling jacket prevents the temperature rising due to the respiration of *Penicillium*: it keeps the medium at different optimum temperatures throughout the production process. Everything that enters the fermenter must be sterilised: if the medium becomes infected by another microorganism, or if the *Penicillium* species changes (mutates), the process is stopped, the spoiled culture is destroyed and everything has to be sterilised again.

Finally penicillin has to be extracted in pure form from the culture. This is made easier by the fact that *Penicillium* secretes penicillin into the medium: it does not have to be extracted from the hyphae of the fungus. When certain potassium compounds are added to the culture, the penicillin forms crystals which sink to the bottom of the culture where they can be collected.

(b) Insulin production

Section 25.4 describes how bacteria called *Escherichia coli* have been successfully programmed to produce the human hormone **insulin** by **genetic engineering**. Genetically engineered cultures of *E. coli* are now being grown in fermenters to produce human insulin. In the future, genetically engineered bacteria may be grown in fermenters to produce a wide range of valuable biological products.

Figure 7.5   A penicillin fermenter

(c) Single-cell protein

Units 6 and 7 describe biotechnology in which microorganisms produce things for us to eat. We can eat microorganisms themselves: the food called **single-cell protein** comes from bacteria or single-celled plant-like protoctists; the food called **mycoprotein** comes from fungi consisting of threadlike hyphae (*mykes* is the Greek for *fungus*). Some people call it all 'bug-grub'.

Section 1.6 describes how Mexican Aztecs were growing and harvesting a plant-like microorganism called *Spirulina* as a green ooze on the surface of freshwater lakes four hundred years ago. Although *Spirulina* is a spiral thread, every cell in it lives independently and it is therefore single-cell protein. People of Chad in Africa have also been growing and eating it. Because it is plant-like, it does not need organic food: it can make its own organic food using the energy from sunlight. The substances it does need are simple, cheap and inorganic. *Spirulina* seems to have real advantages as a food for people in developing countries because it

○ has an annual yield in mass per hectare ten times that of wheat
○ has a high protein content
○ uses renewable energy (sunlight)
○ can be grown and collected from the surfaces of natural or artificial lakes without expensive raw materials or machinery
○ can be dried in the sun
○ can be given different flavours according to local tastes

The excess yeast cells produced by the beer breweries are another form of single-cell protein that has been added to other foods and eaten for many years.

Because they are cheap, manufacturers like to use waste products of other industries as a food source for microorganisms which can produce single-cell protein. Organic waste products which have been used successfully to produce single-cell protein include whey from cheese production, starch from the processing of potatoes and methane gas released in oil fields. Sometimes the single-cell-protein manufacturer even gets paid to take the waste product away. The manufacturer

Figure 7.6 Mycoprotein

who takes sulphite waste away from the paper industry is an example. Because sulphite waste causes serious pollution of rivers, the paper industry is glad to pay to have it taken away. But it is far from useless. Certain yeasts grow well on it forming single-cell protein used in animal foods. The USSR hopes to provide protein for all their animal food from yeasts by the end of this century.

There has been a vast investment of money in new factories and research to produce attractive food for humans from single-cell protein. Mycoprotein, shown on the fork in Figure 7.6, is high in protein (47 per cent) and low in fat (14 per cent). People enjoy eating it because, like meat, it has a fibrous structure. It also helps that it is grown on potato starch: people can be put off by food which they are told is grown on methane gas or sulphite waste.

Questions

Q 7.1. Copy and complete the table.

	One example	Source of food	Importance for humans
Saprophyte Parasite			

Q 7.2. Give the one-word answer that best fits each of the following descriptions:
(a) microorganisms that cause diseases
(b) microorganisms that digest dead organic matter externally
(c) building up new substances from simpler raw materials within the cells of living organisms.
(d) a substance produced by a microorganism which stops the growth of another microorganism.

Q 7.3. Look at the fermenter in Figure 7.4 and explain the function of (a) the stirrer, (b) the acid or alkali reservoir, (c) the water jacket, (d) the air filter and pump and (e) the sampling tube.

Practical work

Experiment 7.1 To test the hypothesis that fungi secrete enzymes on to their food
Note to teachers
Before allowing students to carry out this experiment you should read *Microbiology: An HMI guide for schools and non-advanced further education* (HMSO, 1985).

Students need to know the iodine test for starch which is described in Experiment 15.1.

This experiment lasts at least a week. A practical class is needed for students to inoculate the agar plates with *Rhizopus*. One week later another practical class is needed for them to flood the agar plates with iodine solution. The *Rhizopus* cultures should be incubated at 25°C between the two practical classes.

Strict supervision is needed when students open the Petri dish containing the stock culture of *Rhizopus* to ensure that they do not risk releasing spores into the air. You should demonstrate the way to open the dish as little as possible (see Figure 2.12) or you may prefer to inoculate the agar yourself.

Pure cultures of *Rhizopus* can be obtained from biological suppliers.

A day or two before the practical class, make the starch agar by mixing 1 g of agar and 0.3 g of starch in 100 cm³ of distilled water in a Universal bottle. Autoclave this mixture, with the top loosely screwed in place, at a gauge pressure of 100 kPa (15 lb in⁻²) for 15 minutes. Aseptically pour the agar to cover the bottoms of sterile Petri dishes to an even depth of about 5 mm: 100 cm³ of agar will be enough for about five Petri dishes; each student needs one Petri dish of starch agar. The agar will set in about 15 minutes. The Petri dishes should then be left upside down and open to dry the surface of the agar. Store them in a refrigerator.

Sodium-chlorate(I) solution deteriorates during storage and should be freshly made before this practical.

Students can combine this experiment with Experiment 2.1 where they measure the diameter of the *Rhizopus* colony at daily intervals as an indication of its growth.

Materials needed by each student (Practical Class 1)
1 sterile Petri dish of starch agar
1 inoculating loop
1 Bunsen burner
1 spirit marker
adhesive tape
about 100 cm³ 1% sodium-chlorate(I) solution to deal with spillages

Materials needed by the class (Practical Class 1)
an incubator at 25°C
hand-washing facilities in the laboratory
an active agar-based culture of *Rhizopus* in a Petri dish

Instructions to students (Practical Class 1)
1. Your Petri dish is sterile. It is important to keep it sterile. Open it only when these instructions say you should. Even then, open it as little as possible, keeping the lid down on one side so that it opens at an angle, and do not breathe on it (see Figure 2.12). Treat the Petri dish containing the culture of *Rhizopus* in the same way: you must avoid contaminating it and you must also avoid releasing spores into the air.
2. Turn your Petri dish upside down on its lid. Use the spirit marker to write *Rhizopus* (the name of the fungus to be grown), your initials and

the date near the edge of the dish. Turn the dish the right way up and leave it on the bench.

3. Sterilise the inoculating loop (see Figure 2.13) by holding it in a blue Bunsen flame until the wire at its end glows. Take the wire out of the Bunsen flame and let it cool without touching anything.

4. Go to the Petri dish containing the culture of *Rhizopus*. Lift its lid so that you can just get the sterile inoculating loop on to the *Rhizopus*. Use the loop to remove a small piece of the *Rhizopus* from the edge of the colony. Immediately close the Petri dish containing the culture and lift the lid of your own Petri dish just enough to allow you to put the *Rhizopus* on the agar in the middle of your Petri dish. Immediately withdraw the loop and close the lid.

> **If you drop any *Rhizopus*, pour some of the sodium-chlorate(I) solution over it and inform your teacher.**

5. Put the wire of the inoculating loop back into the Bunsen flame until it glows and then put it down to cool.

6. Stick two short pieces of adhesive tape across the edge of the dish to fix the lid to the base. Do not seal the lid all the way round the dish.

7. Put your dish, base uppermost, in the incubator at $25°C$. Wash your hands before leaving the laboratory.

Materials needed by each student (Practical Class 2)
1 spirit marker
10 cm^3 of iodine in potassium-iodide solution
the incubated culture of *Rhizopus* from the previous practical class

Instructions to students (Practical Class 2)
1. Use the spirit marker and a ruler to draw two lines at right angles to each other across the middle of the base of the Petri dish. Label the ends of these lines A, B, C and D.

2. Look at the surface of the *Rhizopus* colony through the lid of the Petri dish. Make a drawing of the Petri dish showing how the *Rhizopus* colony has grown over the surface of the starch agar. Add the labelled lines which you marked on the Petri dish to your drawing to use for reference.

3. Remove the Petri-dish lid and pour iodine solution over the whole surface of the starch agar. After a few minutes wash the iodine solution away and examine the agar surface against a white background.

4. Make another drawing of the Petri dish showing the position of the blue-black colour. Add the labelled reference lines to your drawing.

Interpretation of results
1. Compare the positions of the blue-black colour and the growth of *Rhizopus* in your drawings. Is there any relationship between the two?

2. Why did some parts of the starch agar turn blue-black with iodine solution? Why did other parts of the starch agar not turn blue-black?

3. What do your results suggest about the release of enzymes from *Rhizopus* and about the movement of the enzymes through the agar?
4. In Practical Class 1 you inoculated starch agar with a small piece of the *Rhizopus* culture. Suggest why you were told to take this piece from the edge of the old colony.

NUTRITION IN FLOWERING PLANTS

8.1 Introduction

Flowering plants, like all living organisms need a supply of food. They need it as a source of energy in respiration and they need it as raw material for growth and repair. Animals and most microorganisms get their food in an organic form: they eat other organisms or the products of other organisms (such as fruit and eggs) or, nowadays, the organic substances made in laboratories and factories. Animals and the microorganisms that do this are called **consumers**.

Because flowering plants can make their own organic food from simple inorganic substances and an outside source of energy, they are called **producers**. (See Section 10.4 for the few bacteria that can also do this.) Once the producers have made their food they use it in the same way as the consumers do: as a source of energy and as raw material for growth and repair.

8.2 Photosynthesis

The simple inorganic substances from which flowering plants make their food are **carbon dioxide** and **water**. These contain no energy that a flowering plant can use: an outside source of energy is needed to combine them into a compound that the plant can use as food. The source of energy is **light**; the food compound that is made is the simple sugar, **glucose**, and the waste product that is left over is **oxygen**. This chemical reaction, called **photosynthesis**, can be shown in words as

carbon dioxide + water + light energy → glucose + oxygen

and in chemical symbols as

$$6CO_2 + 6H_2O + \text{light energy} \rightarrow C_6H_{12}O_6 + 6O_2$$

Light energy is trapped by photosynthesis and converted into chemical energy in the compound glucose.

(a) Carbon dioxide
Carbon dioxide is a gas which is present in air only in small amounts: about 0.04 per cent of air is carbon dioxide. But carbon dioxide is continually added to air by the respiration of all living organisms and by the burning of fuel: wood, coal, gas, oil and petrol all give off carbon dioxide when they burn. There is no

95

danger that carbon dioxide will run out: in fact it is slowly increasing in the air because so much burning now takes place. Carbon dioxide dissolves in water, where there are also dissolved hydrogencarbonate ions which can release carbon dioxide: flowering plants living in water therefore also have a supply of carbon dioxide.

(b) Water

Flowering plants that live on land get the water for photosynthesis through their roots from water in the soil. The water travels through the plant in veins called **vascular bundles**. In Experiment 9.1 you will be able to see the path that water takes as it travels through the plant. Plants that live in water get it from their surroundings.

(c) Energy

Light rays are a form of energy: they are wave movements travelling at great speed. Those of a certain wavelength, which are seen by our eyes, are **light rays**. Flowering plants have a green pigment called **chlorophyll** which can absorb some of these light rays and use their energy to build up the simple sugar, glucose, from carbon dioxide and water. The light energy is trapped as chemical energy in the glucose molecule. Plants can use any source of light rays, but the source that does not run out is **sunlight**. Artificial light is used in glasshouses when extra light is needed. In water, only the top few metres get sufficient light for plants to use in photosynthesis.

Figure 8.1 shows how plants, able to detect a source of light (which they need for photosynthesis), grow towards it. These are cress seedlings, grown from seed in a Petri dish, with those on the left receiving light from above and those on the right receiving light from the right-hand side. You can see that the stems of those on the right have grown bent so that all their small leaves face the light.

Other experiments (for example, capping some of the seedlings with foil which will not let light through) show that the tip of the stem is sensitive to the direction of light: capped seedlings will grow straight even if the light comes from the side.

The growth of plants in relation to the direction of light is called **phototropism**. Because the stem grows towards the light, it is said to be **positively phototropic**. Phototropism is an example of a plant responding, in the direction of its growth, to the stimulus of light. Growth responses in plants are always slow. You can

96   Figure 8.1   Growth of cress towards the light

understand what an advantage positive phototropism is to the plant: it puts its leaves into the light for photosynthesis.

8.3 The site of photosynthesis

For photosynthesis to occur, carbon dioxide, water and light energy must come together where there is chlorophyll in the flowering plant. **Leaves**, which are usually broad and flat, are the parts of the plant most suitable for photosynthesis. Leaves
○ contain chlorophyll
○ have a broad flat area which is supported by veins and is exposed to the rays of the sun
○ contain veins (vascular bundles) which supply water
○ are thin, allowing quick diffusion of carbon dioxide to all parts inside the leaf
The site of photosynthesis is in the cells inside the leaf. Figure 8.2 shows two views of the sort of cell that carries out photosynthesis in a leaf. The living parts of the cell, the **cytoplasm** and **nucleus**, are surrounded by a non-living **cell wall** of **cellulose**: the cell wall keeps the cell in shape. The green pigment **chlorophyll** is held in small disc-like shapes, made from specialised parts of the cytoplasm, called **chloroplasts**. The central part of the cell is a **vacuole** containing **cell sap**, which is mainly water. The cytoplasm is surrounded by a **cell-surface membrane**. A **nuclear envelope** surrounds the nucleus.

Figure 8.3 shows the appearance of some green cells of a leaf of the water plant *Elodea*. These cells are surface cells joined to one another without air spaces between them. The cellulose wall can be clearly seen, as can the chloroplasts inside the cells. The chloroplasts appear to be clustered at the sides but this is because you are seeing several chloroplasts at different levels in the deep cytoplasm at the sides of the cell. The chloroplasts you see in the middle of the cell are lying in the thin layers of cytoplasm at the top and bottom of the cell.

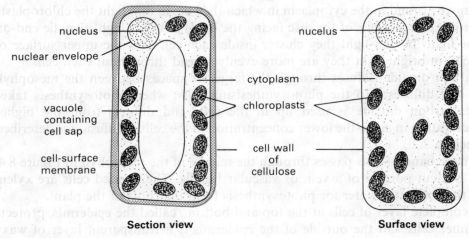

Section view   Surface view

Figure 8.2 Green plant cells

97

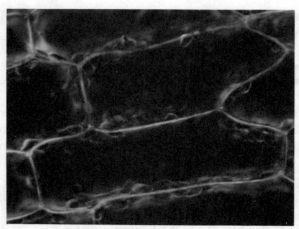

Figure 8.3   Leaf cells of *Elodea*

In Experiment 8.1 you will be looking at solid three-dimensional cells like those in Figure 8.3. You cannot see a nucleus in any of the cells in Figure 8.3: the nucleus is colourless in living cells. Nor can you see the vacuole which is also colourless in most living cells. In Experiment 8.1 you may just be able to see the outlines of the nuclei.

Figure 8.4 shows the cells in a section cut through a plum leaf. You can see how the different parts of the leaf work together to help photosynthesis take place in the cells where there are chloroplasts.

In this leaf the cells that carry out photosynthesis are all on the inside. Between the two outer layers are the cells which contain chloroplasts. These are the cells of the **mesophyll** (*mesophyll* means *middle-leaf*). In the upper part of the mesophyll the cells are closely packed: they look like the cells in Figure 8.2 and contain a lot of chloroplasts. The upper part of the mesophyll gets brighter light from the sun than the lower part: in the lower part the cells are more rounded, there are fewer chloroplasts and there are large air spaces between the cells. Chloroplasts can move around in the cytoplasm in which they lie: in dim light the chloroplasts lie with the broad face of the disc facing the light; in bright light they lie end-on to the light. In dim light they cluster inside each cell near the upper surface of the leaf; in bright light they are more evenly spread throughout each cell.

Carbon dioxide diffuses through the large air spaces between the mesophyll cells and diffuses into the photosynthesising cells: when photosynthesis takes place, carbon dioxide is used up in the cells and diffuses from the higher concentration in air to the lower concentration in the cells. (Diffusion is described in Section 7.2.)

A dark band of cells passes through the middle of the mesophyll in Figure 8.4. This is a cut portion of a vein or vascular bundle. The banded cells are **xylem** cells that bring up water for photosynthesis from the roots of the plant.

A complete layer of cells at the top and bottom, called the **epidermis**, protects the inner cells. On the outside of the epidermis is a transparent layer of waxy material, the **cuticle**, which reduces the amount of water evaporating from the

98

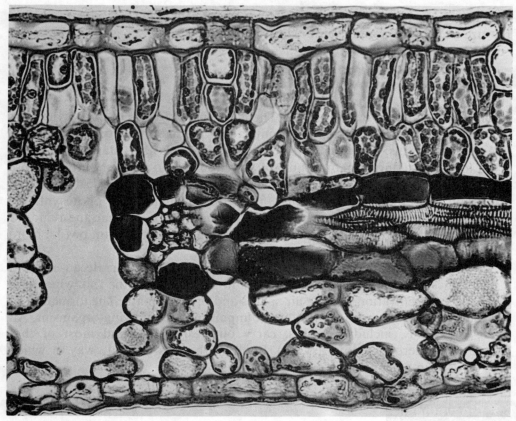

Figure 8.4   Section through a plum leaf

inner cells. It also stops carbon-dioxide gas, needed for photosynthesis, getting into the leaf through most of its surface.

Carbon dioxide gets into the leaf through narrow holes, called **stomata** (singular *stoma*), between special cells on the lower epidermis. You can see the position of a stoma in the lower epidermis in Figure 8.4, but the hole is not very clear. Figure 8.5 is a close-up view of a single stoma in the lower epidermis. The two

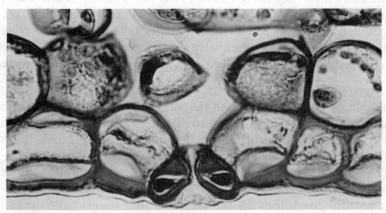

Figure 8.5   Stoma between two guard cells   99

cells on either side of the stoma are called **guard cells**. They are able to change the size of the stoma and even to close it.

8.4 The products of photosynthesis

Look at the equation in Section 8.2. There are two products of photosynthesis: glucose and oxygen. They cannot accumulate in the mesophyll cells indefinitely.

As glucose forms, it is usually stored in the chloroplasts as **starch**. If you test a leaf for glucose during photosynthesis, you are likely to find very little because it is changed to starch so quickly after it is formed. If you test a leaf for starch during photosynthesis, you find quite a lot. But at night, when photosynthesis stops because there is no light, the starch is changed back to sugar and removed from the leaf through the veins either to a growing part of the plant or to a more permanent storage place.

Throughout photosynthesis waste oxygen is given off as a gas. Because it is in a high concentration in the cells, it diffuses to the outside of the leaf, where it is in a lower concentration, via the air spaces and open stomata. The formation of oxygen during photosynthesis is of great importance to all living organisms. All living organisms, including plants, carry out respiration, and most of that respiration is aerobic (using oxygen). Photosynthesis is the only way in which oxygen used in respiration is replaced in the air.

8.5 Biosynthesis

Section 7.6 deals with biosynthesis in microorganisms. Biosynthesis is the use of raw materials and energy to build up new substances within the cells. All living organisms are biosynthesising when they grow and repair themselves. Photosynthesis is one example of biosynthesis in plants: by photosynthesis plants make glucose from carbon dioxide, water and light energy.

Plants need other substances besides glucose to make most of the organic compounds of which they are formed. These other substances are all provided in the form of simple inorganic ions dissolved in water, which land plants take in from the soil by their roots and water plants take in from their surroundings. Two examples of the many elements which are essential for biosynthesis in plants, and which they get as ions dissolved in water, are **nitrogen** and **magnesium**.

(a) Nitrogen
Flowering plants absorb nitrogen in the form of simple inorganic **ammonium** ions and **nitrate** ions. They need nitrogen to add to the substances made from glucose in order to make amino acids. From amino acids they build up proteins (see Section 15.4). They also need nitrogen to make new DNA molecules. Early signs that nitrogen is lacking in a plant are poor growth and yellowing of old leaves which die and drop off.

Under natural conditions all the elements that plants take up as ions, which they build into their organic compounds, are returned to the soil when the plants die: decay by saprophytes breaks down the organic compounds to simple inorganic ions. When we grow crops, we take away parts of them or even all of them with the result that many or all the elements are not returned to the soil. This is why it is vital, when crops are grown, that humans should replace these elements. Farmers and gardeners replace them by adding either **organic** or **inorganic fertilisers**.

(i) Organic fertilisers

A farmer spreading **dung**, the faeces of animals, and a gardener spreading **compost**, partly decomposed plant material, are adding organic material to the soil. They are giving the soil the elements it would have received naturally if humans had not interfered in the first place by harvesting the crops. In the soil, saprophytes break down the organic compounds in these fertilisers to simple inorganic ions.

(ii) Inorganic fertilisers

Crops in the UK are now grown so intensively, with increasing yields from the same plot of ground year after year, that there are no longer sufficient sources of organic fertilisers to replace the elements taken away when crops are harvested. Even people gardening on a small scale may not be able to get enough organic fertilisers for their needs. A vast industry now exists to manufacture and supply inorganic fertilisers containing such ions as ammonium to provide the nitrogen that all plants need. Such vast amounts of inorganic fertilisers have created serious pollution of water (see Section 13.4).

(b) Magnesium

The element magnesium forms part of the chlorophyll molecule and is vital to the plant. Chlorophyll is only a small proportion of a plant's organic compounds but without it the plant cannot photosynthesise, which means it cannot get its food. When crops are harvested, chlorophyll containing magnesium is removed. Fortunately soils rarely lack magnesium ions because the rocks from which soils are made usually contain magnesium salts. A sign that magnesium is lacking is that the older leaves start to lose patches of green colour and the leaves look mottled: this happens because the plant breaks down the chlorophyll in older leaves and takes the magnesium ions to younger growing leaves. Solutions containing magnesium ions can be sprayed directly on leaves or added to soil.

Questions

Q 8.1. Look at the photograph of the section of the leaf in Figure 8.4. Draw an outline plan of this section to show the arrangement of its parts in their correct proportions. Label the cuticle, the upper and lower epidermis, the mesophyll, the position of a stoma, and the vascular bundle and xylem. Do not draw separate cells.

Q 8.2. Look at the photograph of the section of the leaf in Figure 8.4. Choose a single mesophyll cell in which you can see a nucleus. Draw an enlarged version of that cell to show as accurately as you can its shape, its nucleus, the position and number of its chloroplasts and the thickness of its wall. Label the parts you have drawn.

Q 8.3. Design an experiment which you could carry out yourself to test the hypothesis that plants grow better if they are given some inorganic fertiliser.

*Experiment 8.1 To investigate the structure of an **Elodea** leaf with a light microscope*
Note to teachers
You will need live *Elodea* plants for Experiment 8.4 as well as for this one. In this one students can make their own temporary wet mounts. *Elodea* can be bought from most aquarium suppliers and grown in an indoor aquarium or outdoor pond. It is also fairly common in rivers and ponds in the UK. Outdoors, *Elodea* stops growing and dies back in winter. Prepared slides of *Elodea* leaves can be bought from biological suppliers.

The LEAG Series 17 Syllabus requires every student to produce one accurate microscope drawing with a scale. Under high power a specimen (such as a cell) can be measured only when a micrometer graticule is first used on the microscope stage to calibrate an eyepiece graticule: such calibration is too difficult for GCSE students. But under low power they can use a micrometer graticule to make a direct measurement of a microscope specimen (such as an *Elodea* leaf). This is the only Experiment in the Syllabus that lends itself to the use of a micrometer graticule by GCSE students and hence the only one that enables them to produce an accurate microscope drawing with a scale.

Materials needed by each student
1 dropper pipette
1 glass microscope slide and coverslip
1 micrometer graticule
1 pair of forceps
1 scalpel
1 mounted needle
1 microscope
1 piece of fresh *Elodea*

Instructions to students
1. If you have been given a ready-made preparation of *Elodea* leaf, go straight to instruction 5.
2. Put two drops of water on the microscope slide.
3. Use the forceps and scalpel to remove one leaf from the *Elodea* plant. Put it in the drop of water on the microscope slide.
4. Hold the cover slip at an angle at the edge of the drop of water on the slide. Put the mounted needle under the top edge of the cover slip and gently lower it, like a hinged trapdoor, on to the leaf. If you do this carefully, you will not trap any air bubbles under the coverslip.
5. Put the microscope on a firm surface. Depending on the type of microscope you have, either switch on the built-in light or adjust the mirror to reflect as

much light as possible into the microscope. Turn the nosepiece of the microscope so that the low-power objective lens clicks firmly into place.

6. Clip the slide on the stage of the microscope so that you can see the light shining through the leaf.
7. Turn the coarse-focus screw of the microscope until the bottom of the objective lens is as close to the slide as it will go. So long as you are using the lowest-power objective lens there is no danger that it will hit the slide. Looking down the microscope, turn the coarse-focus screw until parts of the leaf come into focus. Continue to turn the coarse-focus screw slowly and you will notice that, as some parts of the leaf go out of focus, others come into focus. This is because the leaf has several layers of cells and the microscope can focus on only one layer. Make a large accurate outline drawing of the leaf.
8. Place the micrometer graticule on the coverslip so that the calibrated part of it lies across the width of the leaf. The graticule is like your ruler except that it is only 10 mm (1 cm) long and is numbered in 1 mm intervals subdivided into 0.1 mm intervals. Looking down the microscope, turn the coarse-focus screw until the graticule is in focus. The leaf will still be visible but will be slightly out of focus. Use the graticule to measure the widest part of the leaf and record it at the side of your drawing.
9. By moving the slide about and focusing on different layers of cells, see how many different types of cell you can find in the leaf. If you want to look at any cell at higher magnification, first move the slide until the cell is directly in the middle of your field of view, then turn the nosepiece of the microscope until a higher-power objective lens clicks firmly into place. You must take great care not to let this objective lens touch the slide. While looking at the objective lens from the side, turn the coarse-focus screw of the microscope until the bottom of the objective lens is as close to the slide as it will go. When you look down the microscope at this magnification, *always* rack upwards. Now you can re-examine the cells.
10. Make a large accurate drawing of each type of cell you see. Appendix C tells you how to draw biological material.

Interpretation of results
1. Most of the cells you saw contained large disc-like objects in the cytoplasm. What were these objects?
2. Label your drawing to identify these objects and any others from the list below which you were able to see:

> cell wall
> cytoplasm
> nucleus
> vacuole

3. If you were looking at a living leaf, you may have seen movement in the cytoplasm of the *Elodea* cells. If you did, describe the movement and suggest why it occurs.

103

Experiment 8.2 To test the hypothesis that plants need light to produce starch

Note to teachers

Potted Busy Lizzie plants (*Impatiens wallerana*), also known as Patient Lucy, can be bought from garden centres and can easily be propagated in a laboratory: they can be used in this experiment and in Experiments 8.3 and 9.1. Potted geraniums (*Pelargonium*) are also suitable and students may find their broad leaves easier to handle.

You must destarch the plants before they can be used. To do this, water them and leave them in complete darkness for at least 48 hours. If you do not have a suitable cupboard, you can put the potted plants inside a heavy-duty black dustbin liner and tie the top. Test a leaf from one of the plants for starch before the practical class begins. If starch is present, continue to destarch the plants and postpone the experiment. Depending on the number of suitable leaves, one plant is enough for about four students.

After covering a leaf with aluminium foil, students need to leave the plant to photosynthesise before testing the leaf for starch. If students put the plants under a strong bench lamp, starch will be found within one hour. About one and a half hours is therefore needed for this experiment. If the class does not last as long as this, students should leave the plants in bright light until the next day (when the 'Instructions to students' will need slight modification).

Bench lamps with incandescent bulbs can be used in this experiment, but they make the leaves very hot. A fluorescent strip, held horizontally by clamps, boss heads and stands, gives bright light without excessive heat.

Students need to know the iodine test for starch which is described in Experiment 15.1.

For the safety of students it is advisable for you to dispense the ethanol from one point away from lighted Bunsen burners.

Several students should work together in a group.

Materials needed by each student

1 250 cm^3 beaker
1 test-tube rack with 1 test-tube
1 pair of forceps
1 plain white tile
1 Bunsen burner, tripod and gauze
1 spirit marker
1 1 cm × 12 cm strip of aluminium cooking foil
5 cm^3 iodine in potassium-iodide solution

Materials needed by the class

104 ethanol (about 10 cm^3 per student to be dispensed by the teacher)

1 fluorescent light strip, supported by clamps, boss heads and stands
healthy destarched potted plants (1 per four students)

Instructions to students

1. Use the spirit marker to write your initials on the piece of aluminium foil.
2. Choose a healthy-looking leaf towards the top of the potted plant. Without harming it, wrap the aluminium foil around both top and bottom surfaces of part of the leaf so that no light can get to them.
3. When all the members of your group have put aluminium foil around leaves of the plant, put the plant under the light so that it is nearly touching the leaves.
4. Water the soil if it is dry.
5. After about an hour half fill the beaker with water and boil it over the Bunsen burner, tripod and gauze. Take your labelled leaf from the plant and remove the aluminium foil. Use the forceps to hold it in the boiling water for a few seconds.
6. TURN THE BUNSEN BURNER OUT.
7. Use the forceps to push the leaf carefully to the bottom of the test-tube. Take the test-tube to your teacher, who will cover the leaf with ethanol.
8. Put the test-tube into the beaker of hot water and leave it to boil for a few minutes until the colour has gone from the leaf. DO NOT RELIGHT THE BUNSEN BURNER.
9. Use the forceps to remove the leaf from the test-tube and dip it in the hot water in the beaker to wash it.
10. Spread the leaf on the white tile and cover it with iodine solution.
11. After about a minute hold the tile under a running tap to wash away the iodine solution.
12. Draw the leaf to show the positions of the white and blue-black areas.

Interpretation of results

1. What substance turns blue-black with iodine solution?
2. How does a plant make this substance?
3. Explain why only some parts of your leaf turned blue-black with iodine solution.
4. Why were you told to
   (a) cover only part of the leaf with aluminium foil (instruction 2)?
   (b) water the soil if it was dry (instruction 4)?
   (c) boil the leaf in water for a few seconds (instruction 5)?
   (d) turn the Bunsen burner out (instruction 6)?
5. Before the experiment started, your teacher had left all the plants in complete darkness for 48 hours. Explain why this was important.

Experiment 8.3 To test the hypothesis that a photosynthesising green leaf takes up carbon dioxide

Note to teachers

Potted Busy Lizzie plants (*Impatiens wallerana*), also known as Patient Lucy, can be bought from garden centres and can easily be propagated in a laboratory: they can be used in this experiment and in Experiments 8.2 and 9.1. Potted geraniums (*Pelargonium*) are also suitable.

You must destarch the plants before they can be used. To do this, water them and leave them in complete darkness for at least 48 hours. If you do not have a suitable cupboard, you can put the potted plants inside a heavy-duty black dustbin liner and tie the top. Test a leaf for starch before the practical class. If starch is present, continue to destarch the plants and postpone the experiment.

The experiment lasts for two practical classes. The plants must be left undisturbed in a well lit place for up to a week between the two practical classes.

Students should have done Experiment 8.2 before they do this one. They need to know the iodine test for starch which is described in Experiment 15.1.

Several students should work together in a group in the first practical class.

> **For the safety of students it is advisable for you to dispense the ethanol from one point away from lighted Bunsen burners during the second practical class.**

Materials needed by each group of students (Practical Class 1)
2 5 cm diameter Petri dishes
1 spirit marker
2 transparent polythene bags with twist-tie fasteners, large enough to contain the potted plants
15 g soda lime
15 g chalk granules
2 healthy destarched potted plants

Instructions to students (Practical Class 1)
1. Label the plant pots A and B and add your initials. Water the soil if it is dry.
2. Put about 15 g of soda lime in one of the small Petri dishes. Put the dish on the surface of the soil in pot A. Put about 15 g of chalk granules in the other small Petri dish and put it on the surface of the soil in pot B.
3. Check the polythene bags for holes by holding the opening over your mouth and blowing into them. If any of your breath leaks from the bag, use another. Put each potted plant in a polythene bag. Close the top of each bag securely with a twist tie so that no air can get in.
4. Leave the two potted plants in their polythene bags where they will be well lit and undisturbed until your next practical class.

Materials needed by each student (Practical Class 2)
1 250 cm^3 beaker
1 test-tube rack with 2 test-tubes
1 pair of forceps
1 plain white tile
1 Bunsen burner, tripod and gauze
1 spirit marker
106   10 cm^3 iodine in potassium-iodide solution

Materials needed by the class
ethanol (about 20 cm^3 per student to be dispensed by the teacher)

Instructions to students (Practical Class 2)
1. Half fill the beaker with water and boil it over the Bunsen burner, tripod and gauze. While you are waiting for the water to boil, label two test-tubes A and B.
2. Take a leaf from plant A and use the forceps to hold it in the boiling water for a few seconds. Use the forceps to push the leaf carefully to the bottom of the test-tube labelled A. Remove a leaf from plant B and treat it in the same way as the first leaf before putting it in the test-tube labelled B.
3. TURN THE BUNSEN BURNER OUT.
4. Take both test-tubes to your teacher, who will cover both leaves with ethanol.
5. Put the test-tubes into the beaker of hot water and leave them to boil for a few minutes until the colour has gone from both leaves. DO NOT RELIGHT THE BUNSEN BURNER.
6. Use the forceps to remove the leaf from test-tube A and dip it in the hot water in the beaker to wash it.
7. Spread the leaf on the white tile and cover it with iodine solution.
8. After about a minute hold the tile under a running tap to wash away the iodine solution. Record the colour of the leaf.
9. Use the forceps to remove the leaf from test-tube B and dip it in the hot water in the beaker to wash it.
10. Repeat instructions 7 and 8 with the leaf from test-tube B.

Interpretation of results
1. What substance turns blue-black with iodine solution?
2. How does a plant make this substance?
3. Soda lime absorbs carbon dioxide from the air, while chalk granules have no effect on the carbon dioxide in the air. Use this information to interpret the results of your experiment.
4. What was the purpose of plant B in this experiment? Would your conclusion have been different if it had not been present?
5. Gardeners often put plant cuttings in pots and tie them inside polythene bags as you did with plants A and B. Before tying the top of the bag they often breathe into the bag.
   Suggest why these gardeners
   (a) put the planted cuttings in polythene bags
   (b) breathe into the bag before tying the top

Experiment 8.4   To test the hypothesis that a submerged water plant gives off oxygen
Note to teachers
You will need live *Elodea* plants for Experiment 8.1 as well as for this one. *Elodea* can be bought from most aquarium suppliers and grown in an indoor aquarium or outdoor pond. It is also fairly common in rivers and ponds in the U.K. Outdoors, *Elodea* stops growing and dies back in winter.

Make sure that *Elodea* is actively photosynthesising before the practical class   107 begins. This can be done by keeping the plants well lit and, if necessary, adding a spatula end of potassium hydrogencarbonate just before the experiment.

The experiment lasts for two practical classes, one in which the students set up the materials and another in which they investigate the gas produced during photosynthesis. The materials set up must be left undisturbed in a well lit place for up to a week between the two practical classes.

Materials needed by each student (Practical Class 1)
1 500 cm^3 glass beaker
1 short-stemmed glass filter funnel
1 1 cm^3 pipette fitted with a pipette filler
1 boiling-tube rack with 1 boiling tube
1 pair of forceps
1 pair of scissors
1 spirit marker
sprigs of active photosynthesising *Elodea*

Instructions to students (Practical Class 1)
1. Cut about 10 cm from a sprig of the pond weed which is producing bubbles of gas and put it in the beaker. Half fill the beaker with water.
2. Put the filter funnel over the piece of pond weed in the beaker and add more water until it is about 1 cm above the tip of the spout of the funnel.
3. Leave the apparatus in a well lit place while you follow instruction 4. If at the end of this time the pond weed no longer produces bubbles of gas, replace it with another piece.
4. Pipette 1 cm^3 of water into the boiling tube and mark its position on the outside of the tube. Pipette another 1 cm^3 of water and mark its position on the outside of the tube. Carry on doing this until the tube is completely full of water.
5. Put your thumb over the end of the boiling tube and quickly turn it upside down with its open end below the water in the beaker. Remove your thumb from the tube and carefully lift the tube over the tip of the funnel. Lower it on the funnel so that it rests upside down. If you have lost a large volume of water in doing this, refill the tube with water and try again. Use the marks you made on the tube to estimate the volume of any air space in the tube. Record the volume of the air space if there is one.
6. Write your initials and the date on the beaker. Leave the beaker and its contents where they will be well lit and undisturbed until your next practical class.

Materials needed by each student (Practical Class 2)
1 wooden splint

Instructions to students (Practical Class 2)
1. Record the new volume of the gas inside the tube.
2. Light a wooden splint and let it burn for a few seconds before blowing it out again. Hold this glowing splint in one hand and remove the tube from the beaker of water with the other. Let the water run out of the tube as you take

108

it out of the beaker. Put the glowing splint in the tube of gas and record what happens.

Interpretation of results
1. What was the change in the volume of gas in the tube?
2. What did your test with the glowing splint show about the gas which collected in the tube?
3. Explain the change in the volume and composition of gas in the tube.
4. Did all the members of your class obtain similar results? If so, does this make your conclusions more reliable? Explain your answer.
5. Design a control experiment which could be performed to show that your interpretation is valid. (Appendix B tells you more about a control experiment.)
6. Suggest a way in which this experiment could be modified to test the hypothesis that carbon dioxide is taken up by a submerged *Elodea* plant in the form of hydrogencarbonate ions. You can increase the concentration of these ions by adding potassium hydrogencarbonate to the water. This is a suitable experiment for an Individual Study.

FLOWERING PLANTS

9.1 Structure

A flowering plant is made up of millions of cells. The cells are organised into **tissues**, collections of cells that look more or less alike and have the same functions (for example, mesophyll tissue). The tissues are joined to one another to form **organs**, such as leaves and stems, which carry out the major living processes of the plant. Organs together form the whole flowering plant. All large **organisms** are made up of organs, which are made up of tissues, which are made up of cells.

The main organs of a flowering plant are the roots, stems and leaves and the parts concerned with reproduction: sepals, petals, stamens and carpels. Figure 9.1 shows three different examples of flowering plants. Stems, leaves, flowers and the small **buds** that protect these parts when they are young together form a **shoot**.

(a) Roots

When a seed grows, the first part to appear is a young root. This becomes the main root or **tap root** of the plant. In some plants, such as carrots and the parsley shown in Figure 9.1, the tap root remains the main root throughout life. In other plants, such as cereals and the mint and chive shown in Figure 9.1, the tap root is replaced by **fibrous roots** that grow out from the stem.

The functions of the roots of flowering plants growing on land are
○ to hold the plant in the soil
○ to absorb water from the soil
○ to absorb ions in solution from the soil
In many plants, such as carrots and mooli, the roots also store food.

(i) Water

Water is vitally important to flowering plants. They need it as a raw material for photosynthesis. They need it also in cytoplasm, nuclei and the cell sap in the vacuoles.

A cell that is full of water has its cytoplasm pressed up against the cellulose cell wall. When all the cells are in this state, the tissue is firm (turgid): if you cut a 0.5 cm slice of raw potato, it is firm at first, but after a few hours in dry air, it will feel flabby. The firmness given by water helps the stem to support leaves and flowers in the air: cut flowers soon droop if the stems are not put in water.

The cytoplasm in the cell is mainly water: here all the chemical reactions that take place in a living cell are carried out in solution. Diffusion within the cells takes place in water and ions pass into the roots of plants in solution (dissolved) in water.

Water is continually lost from the surfaces of the plant which are exposed to 111

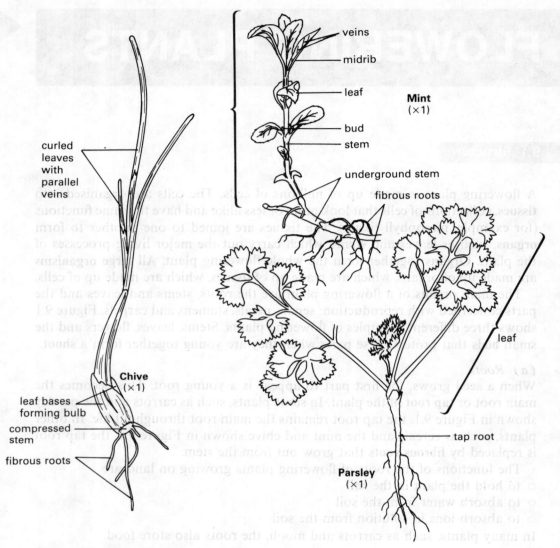

veins
midrib
leaf

Mint
(×1)

bud
stem

underground stem

fibrous roots

curled
leaves
with
parallel
veins

leaf

Chive
(×1)

leaf bases
forming bulb

compressed
stem

fibrous roots

tap root

Parsley
(×1)

Figure 9.1 Flowering plants

air. In particular it evaporates from all the wet cell surfaces into the air spaces
in the mesophyll of the leaves. Because these air spaces are in contact with air
through the stomata, the water vapour diffuses out into the air. This loss of water
in the form of water vapour from the parts of the plant in air is called **transpiration**.
The process takes place in two stages:
○ evaporation of water into the air spaces of the mesophyll
○ diffusion of water vapour out via the stomata
Transpiration is important because it provides the means by which water,
containing dissolved ions, moves from the roots through the stem to the leaves.

When the water evaporates from the mesophyll cells in the leaf, it 'pulls' the
water up the plant. This sounds impossible but you have probably 'pulled' a
drink up through a straw. Because water molecules stay together (by cohesion)

they can be pulled up airtight tubes. A plant's transport system is its **vascular tissue** containing two sets of small tubes. One set of small tubes (the **xylem**) is continuous from the roots through the stems and through the veins in the leaves. In this set of tubes fine columns of water stay unbroken as they are pulled up from the roots to replace the water that evaporates by transpiration from the leaves. In Experiment 9.1 you will be able to find the position of these tubes in the plant.

(b) Stems

Stems are the parts of a flowering plant which hold the leaves and flowers in air. The leaves are held out for photosynthesis and the flowers for reproduction. Vascular tissues extends throughout stems carrying water and dissolved ions from the roots to the leaves. Sugar (formed in photosynthesis) is carried in solution through the other set of tubes (the **phloem**) in the vascular tissue from the leaves to the growing parts of the plant and to the roots and seeds for storage.

Stems are not always above ground. In the mint plant shown in Figure 9.1 you can see that some stems are above ground supporting the leaves and other stems are growing horizontally through the ground. The horizontal stems spread the plant out through the surrounding soil. When older parts of the horizontal stems rot away, they leave separate plants: they are therefore one means by which the plant reproduces and increases in number.

Stems may be so compressed that they are very short. This is shown in Figure 9.1: in the parsley the leaves seem to be growing out of the top of the root; in the chive the stem is surrounded by the swollen bases of the leaves to form a **bulb**.

(c) Leaves

The variety of leaf shape in different flowering plants is enormous. Figure 9.1 shows three different shapes: the mint has a simple outline; the parsley has a leaf deeply divided into a number of segments (called **leaflets**); the chive has a very narrow leaf curled into a tube-like shape. The leaves contain veins which may lie parallel to one another (as in chive) or branch out from a main vein (as in mint and parsley).

The processes taking place in all these leaves are the same:
○ photosynthesis (using energy from sunlight to make glucose)
○ transpiration (resulting in the movement of water and ions through the plant)
○ respiration (releasing energy)
In detail the three processes are
○ carbon dioxide + water + light energy → glucose + oxygen (photosynthesis)
○ the evaporation and diffusion of water (transpiration)
○ glucose + oxygen → carbon dioxide + water + energy (respiration)
These three processes involve a considerable gas exchange between the leaf and the air outside through the stomata.

To work out what gas exchange is taking place between the leaves and the air at any one time you need to remember that
○ photosynthesis takes place only in daylight
○ transpiration takes place only when the stomata are open (and they are usually almost closed at night)

○ respiration takes place all the time

(i) In darkness

Photosynthesis cannot take place at night (except in artificial light) because the energy supplied by sunlight is missing. Respiration takes place as usual in the cells of the leaf: oxygen is used up and carbon dioxide is produced. This leads to concentration gradients and, provided the stomata are not completely closed, as they may be at night, a small amount of oxygen diffuses into the leaf and a small amount of carbon dioxide diffuses out of the leaf.

Only small amounts of these gases are involved because respiration in a flowering plant is never vigorous. Plants need energy to keep the cells alive and to make the complex compounds they need for growth and repair (biosynthesis), but they do not need the large supplies of energy which animals need for movement and which warm-blooded animals need to keep their temperature up.

Transpiration takes place in darkness if the stomata are slightly open and if the air outside is not 100 per cent saturated. Water vapour collects in the air spaces of the leaf, which are 100 per cent saturated. This water vapour diffuses out of the leaf provided that the air is drier outside forming a concentration gradient. When the air outside is also 100 per cent saturated, transpiration stops.

(ii) In daylight

Respiration continues in daylight in all the living cells of the leaf. But in daylight photosynthesis in the mesophyll cells (where there is chlorophyll) is much more vigorous than respiration. Since photosynthesis needs carbon dioxide and gives off oxygen, it masks the gas exchange of respiration. Carbon dioxide produced by the cells in respiration never gets outside the leaf because it is immediately used up in photosynthesis. The respiring cells do not get oxygen from outside the leaf because they need use only a small amount of the oxygen produced inside the leaf by photosynthesis.

In daylight the stomata are wide open allowing free diffusion of carbon dioxide into the leaf and of oxygen out of it. Because in daylight the air outside is rarely saturated, transpiration usually occurs freely too: water vapour diffuses out of the leaf along the concentration gradient to the drier air outside. Some water is used by the mesophyll cells in the process of photosynthesis, but this is only about 0.2 per cent of the water absorbed by the plant: almost all the water absorbed by the plant is transpired.

(iii) In half-light

Gas exchanges taking place in the leaf by day and by night are shown in Figure 9.2. With carbon dioxide and oxygen going in opposite directions by day and by night, twice a day between daylight and darkness there is a short time when there is no gas exchange of oxygen and carbon dioxide between the leaf and the air outside: all the oxygen needed for respiration is provided by photosynthesis and all the carbon dioxide given off in respiration is used in photosynthesis. Transpiration will still be taking place unless (which is unlikely at these times) the air outside is 100 per cent saturated.

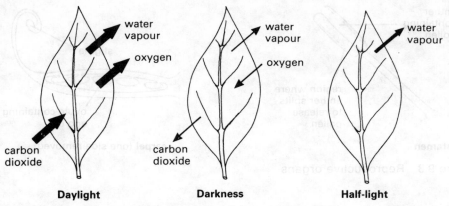

Figure 9.2   Diffusion in daylight, darkness and half-light

Many flowering plants have a short life. Although some, such as trees, live for hundreds of years, others, such as tomato plants, die after only a few months. They must all have a way of **reproducing** themselves. Almost all flowering plants reproduce by **sexual reproduction**, in which two sex cells, called **gametes**, fuse together to form a single cell, called a **zygote**. Fusion of the two gametes is called **fertilisation**. The zygote grows into an **embryo** which is protected inside a **seed**. The embryo grows into a new flowering plant. Section 22.2 gives more details of sexual reproduction.

The plant organs concerned with sexual reproduction are clustered together in the form of **flowers**. Despite the variety in the shape of flowers produced by plants, they all contain the parts that produce the gametes for sexual reproduction. In flowering plants, as in almost all plants and animals, there are two sorts of gamete: **male** and **female**. The female gamete or **egg cell** is usually larger than the male and does not move. The male gamete is smaller, is without a supply of food and does move. In flowering plants the organ that produces the egg cell is called a **carpel** and the organ that indirectly produces the male gamete is called a **stamen**.

The stamens of all flowering plants are similar. Although they do not produce male gametes directly, they produce a powdery substance called **pollen**, which later produces male gametes. Figure 9.3 shows a typical stamen: the powdery pollen is produced in four sacs at the tip; these four sacs form the **anther**.

The carpels of flowering plants, which produce the egg cells, vary greatly. To get some idea of the variety you need only look at different fruits. **Fruits**, which develop from carpels after egg cells have been fertilised, help to protect and spread the seeds that have formed. Oranges, raspberries, kiwi fruits and mangoes have all developed from different carpels. Even cucumbers, tomatoes, chillies and okras have developed from carpels, contain seeds and are strictly fruits. Figure 9.3 shows the carpel of a bean plant: it contains an egg cell in each of the small ovoid bodies called **ovules** and has a special region, the **stigma**, where pollen from the anther must land. After fertilisation, ovules form seeds.

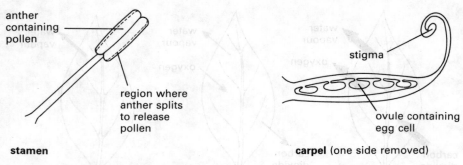

Figure 9.3   Reproductive organs

anther containing pollen

region where anther splits to release pollen

stamen

stigma

ovule containing egg cell

carpel (one side removed)

Petals, the often large and coloured parts of flowers, and **sepals**, the small green parts on the outside of flower buds, are not strictly reproductive organs, but they help in reproduction. Sepals protect the reproductive organs as they are developing; petals also protect the reproductive organs and in many flowers help in the process of pollination.

(a) Pollination

Before fertilisation can take place, the pollen, from which the male gametes develop, must be brought near to the ovule in which the female gamete develops. Pollen is first brought from the anther to the stigma of a carpel: the coming together of pollen and the stigma is **pollination**.

Pollen is carried from the anther to the stigma in different ways. Much pollen is simply carried in the wind (and gives some people hay fever): most or even all of this is wasted but some may land on stigmas of the right species. Insects accidently transfer pollen from anther to stigma when they visit those flowers that form the sugary **nectar** they feed on (see Section 12.2).

Figure 9.4 shows a group of flowers of the oat plant. (The flowers of most cereals and grasses are similar.) They hardly look like flowers because they do not have petals and sepals that you see on many other flowers: they have greeny-brown scales instead. They do have the important reproductive organs, also shown in Figure 9.4, which you can see if you pull a flower open. The carpel in Figure 9.4 contains only a single ovule. Oat-flower pollination (the transfer of pollen from an anther to a stigma) takes place by wind. Pollen blows away from anthers hanging outside the flowers. Some of it may be trapped among the hairs of stigmas also hanging outside the flowers. (It may be trapped among the hairs of stigmas of the same plant or different plants.)

(b) Fertilisation

A pollen grain usually contains two nuclei. One nucleus controls the development of a long **pollen tube** which grows out of the pollen grain when it lands on a suitable stigma. Figure 9.5 is a photograph taken with an electron microscope showing pollen grains and their pollen tubes (at about × 550). The large finger-like processes are the cells on the stigma surface. The photograph is interesting because it shows that different pollen grains have stuck to the stigma surface (look at the different shapes and sizes of the pollen) but that only the round pollen grains of

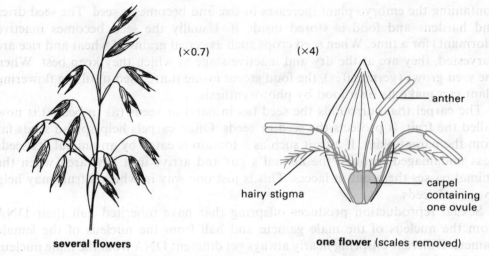

(×0.7)　　　　　　(×4)

anther

hairy stigma

carpel containing one ovule

several flowers　　　　　**one flower** (scales removed)

Figure 9.4　Flowers of the common oat

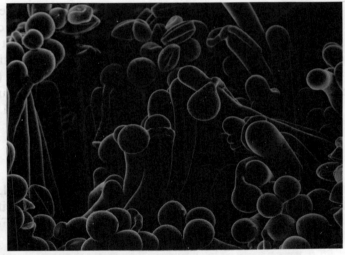

Figure 9.5　Pollen grains and tubes on a stigma

this species have produced pollen tubes. The pollen tubes grow right through the tissues of the carpel into the ovules.

The other nucleus in the pollen grain stays just behind the tip of the pollen tube as it grows through the carpel. When it gets near the egg cell in the ovule, it divides into two to form the nuclei of two male gametes. The male gamete of a flowering plant is not much more than a nucleus. The pollen tube carries the two male nuclei close to the egg cell in the ovule. Fertilisation takes place when one of the male nuclei fuses with the egg cell nucleus to form the zygote. The other male nucleus later helps to provide food for the developing embryo.

(c) Development of seeds and fruits
After fertilisation the zygote divides and grows into an embryo plant. The ovule　117

containing the embryo plant increases in size and becomes a **seed**. The seed dries and hardens and food is stored inside it. Usually the seed becomes inactive (dormant) for a time. When seed crops such as cereal grains of wheat and rice are harvested, they are at the dry and inactive stage at which they keep best. When the seed grows (germinates), the food stores inside it are used until the flowering plant can make its own food by photosynthesis.

The carpel that surrounds the seed (as in oats) or seeds (as in beans) is now called the **fruit**. It protects the seed or seeds. Often carpels help spread seeds far from the parent plant. If a fruit such as a tomato is eaten by an animal, its seeds pass undamaged through the animal's gut and arrive in a new area when the animal passes them out in faeces. This is just one way in which a fruit may help to spread seeds.

Sexual reproduction produces offspring that have inherited half their DNA from the nucleus of the male gamete and half from the nucleus of the female gamete. Different offspring nearly always get different DNA from the male nucleus and different DNA from the female nucleus even when they have the same parents. The offspring of sexual reproduction are therefore nearly always different (and are often very different) from their parents and from one another. They are nearly always different even if, as is possible in plants. the male nucleus and the female nucleus come from the same parent. This **variety** in offspring is of great importance in the **selective breeding** of better plant (and animal) crops (see Section 11.1).

Sexual reproduction is only one method of reproduction. **Asexual** (not sexual) **reproduction** produces offspring that inherit identical DNA from a single parent: this is also of great importance in crop production (see Section 11.1).

Questions

Q 9.1. Draw a table of five columns. Head the columns Roots Stems Leaves Seeds Fruits. Identify as many of the foods in Figure 9.6 as you can. Write the name of each food you identify in what you think is the right column. For example, carrots go in the column headed Roots.

Figure 9.6 Edible plant parts

Q 9.2.
(a) What is transpiration?
(b) How and where does transpiration take place?

118

(c) Why is transpiration of importance to flowering plants?

Q 9.3. Explain why the following are needed by flowering plants: (a) water, (b) nitrate ions and (c) magnesium ions.

Q9.4 Explain the difference between pollination and fertilisation in a flowering plant.

Experiment 9.1 To investigate water transport in a leaf stalk or stem of a plant
Note to teachers
Celery heads are available at supermarkets and greengrocers for most of the year. Each celery head has many leaf stalks.

Potted Busy Lizzie plants (*Impatiens wallerana*), also known as Patient Lucy, can be bought from garden centres and can easily be propagated in a laboratory: they can be used in this experiment and in Experiments 8.2 and 8.3. The movement of water through Busy Lizzie is faster than through celery.

At the end of the experiment students are asked to design further experiments. If time permits, they can be allowed to conduct these experiments, but they will be suitable as Individual Studies only if they test hypotheses and involve suitable measurements.

Materials needed by each student
1 250 cm^3 beaker
1 scalpel
1 stopclock
1 hand-lens
1 ruler graduated in mm
100 cm^3 water coloured with a little methylene blue or eosin
1 fresh leaf stalk of celery or stem of Busy Lizzie

Instructions to students
1. Pour the coloured water into the beaker.
2. Cut about 5 mm from the bottom of the celery leaf stalk or Busy Lizzie stem and stand it in the beaker of coloured water. Start the stopclock.
3. After about 30 minutes stop the clock and remove the plant from the beaker. Rinse it in tap water and measure the length which shows the colour of the dye inside it. Record this distance and the time for which the plant was in the beaker.
4. Cut across the part which is coloured by the dye. Examine the cut end with the hand-lens and find the exact position of the coloured dye. Record this position on a drawing of the cross-section of the part.

Interpretation of results

1. From your data calculate the rate of movement (in mm per minute) of the coloured water by using the formula

$$\frac{\text{distance travelled by the coloured water in mm}}{\text{time in minutes}}$$

2. Compare the result of your calculation with those of others in your class. Was the rate of movement always the same? If not, suggest a reason for the different rates.
3. What plant *tissue* is found in the position of the coloured dye in your drawing? What assumption are you making if you conclude that this is the tissue concerned with water transport? Remember that you started with a mixture of water and dye.
4. Suggest how you could either change this experiment or design another to show the position of water transport tissue in
   (a) a plant leaf
   (b) a plant root.

RECYCLING

10.1 Introduction

Nowadays the word **recycling** suggests piles of waste paper or bottle banks set up by local authorities so that paper and glass may be reprocessed and used again. It may also suggest scrap merchants who buy old car bodies to sell for the metal that can be salvaged from them. Recycling of this kind, which keeps up supplies of scarce raw materials, is a fairly new idea. But recycling has been occurring naturally since the earth was formed four thousand million years ago.

10.2 Water

Water is recycled around us all the time. Rain, snow and hail fall on the earth's surface every day. Some falls on water (on rivers, lakes and sea) or runs into water directly after falling on nearby land. Of the rest of the rain, snow and hail that falls on land, some soon **evaporates** into the air, some soon sinks into the soil and the rest lies for a time as snow or ice before either evaporating or sinking into the soil. Of the water that sinks into the soil, a little is absorbed by the roots of plants: most of this passes straight through the plants and is transpired into the air as water vapour. Most of the water that sinks into the soil goes down to the water table, from which it drains into rivers, lakes and the sea. Some of the water in rivers, lakes and the sea is evaporating all the time.

By one route or another water returns to the air as **water vapour**. In air, water vapour eventually condenses, forming fine droplets of water in clouds, larger droplets of water in rain, or ice crystals in snow or hail. Somewhere rain, snow or hail will again be **precipitated**. The water cycle is summarised in Figure 10.1.

Unfortunately rain, snow and hail may not be precipitated where they are needed by humans and other organisms. Some places get too little water and there is **drought**; others get too much and there is **flooding**. Already we can have a small influence on when and where water falls: in farming areas light aircraft release crystals to precipitate the water vapour in the air sooner than it would precipitate naturally Perhaps one day we shall be able to do away with droughts and flooding.

10.3 Carbon cycle

Carbon is in every organic compound that exists and there is only so much of it on earth. If it were not recycled, there would soon be a shortage of it. It starts   121

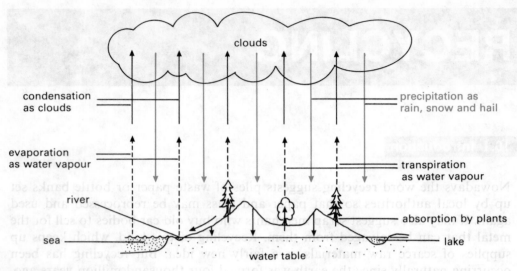

Figure 10.1   Water cycle

to be recycled when a dead body **decays** and is absorbed as food by the saprophytes (which first digest it externally). Inside a saprophyte the carbon compound is either respired as a source of energy, when carbon is given off as carbon dioxide, or it is built up into the body of the saprophyte. The organism that next feeds on the saprophyte may then give off the carbon as carbon dioxide.

If a plant is eaten by an animal, carbon is either returned to the air as carbon dioxide after respiration in the cells or is built up into the body of the animal. In the body of the animal it is either eaten by another animal or used as food (decayed) by a saprophyte after it dies. One way or another, carbon will return to the air as carbon dioxide.

The carbon dioxide in the air will, sooner or later, be absorbed by a plant and be used to build up glucose in photosynthesis. From the chloroplast of a plant, carbon in glucose is either sent to other cells to be used up in respiration (and return to the air as carbon dioxide) or it is built up into one or more of the organic compounds of which a plant is made. The carbon cycle is summarised in Figure 10.2.

Humans interfere with the earth's carbon cycle when they burn carbon compounds (as wood, coal, gas, oil and petrol) and so increase the amount of carbon dioxide in the air. At the same time humans are reducing the number of plants in many parts of the world by cutting down forests and increasing desert lands. The result is that there are not enough plants to remove the carbon dioxide quickly enough.

In the last fifty years the amount of carbon dioxide in air has gone up from 0.03 per cent to 0.04 per cent and, at the present rate of burning, will go up to 0.6 per cent by the year 2050. This may not seem much, but it may have a great effect on the climate: scientists still argue about whether it will lead to a general heating or cooling of the earth.

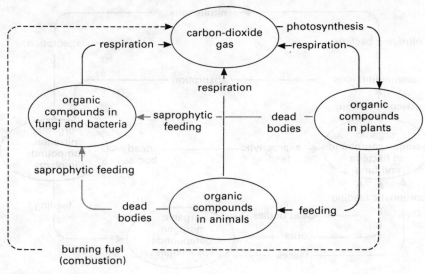

Figure 10.2 Carbon cycle

Nitrogen is just one of the many elements that must be recycled after it has been taken up by plants. Nitrogen is important because it is needed to make every protein and every molecule of DNA in the plant; no plant can grow without it. Although 80 per cent of the air consists of nitrogen gas, flowering plants cannot use it: they have to get their nitrogen in the form of ammonium or nitrate ions.

Unlike carbon (which may be given off as carbon-dioxide gas), organic nitrogen compounds stay in a plant until they are either eaten by an animal or used as food (decayed) by saprophytes. Eaten by an animal (perhaps by an animal that eats the animal that eats it), organic nitrogen may be built into the compounds of an animal or it may be excreted (as urea) or removed as faeces (material that cannot be digested). Organic nitrogen in urea and faeces is also used as food by saprophytes. Therefore, whatever happens to the organic nitrogen compounds in the plant, whether they end up in dead bodies, in urea or in faeces, they all become the food of saprophytes.

Different saprophytes break down the organic nitrogen compounds in stages. They digest them externally, they absorb them and use them as food for growth and repair, and they use them as a source of energy in respiration (including fermentation). Eventually the nitrogen compounds are all broken down (decayed) to release ammonium ions.

Ammonium ions still contain some energy that certain microorganisms can release and use. These are the bacteria mentioned in Sections 7.1 and 8.1. They are called **nitrifying bacteria**. Nitrifying bacteria change ammonium ions to nitrate ions and in the process release energy which they use to make their own food from simple inorganic compounds. This process, called **nitrification**, is like   123

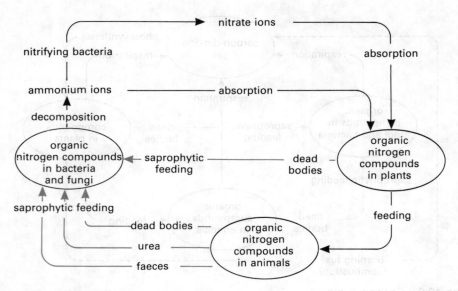

Figure 10.3   Basic nitrogen cycle

photosynthesis except that these bacteria make their food using the energy in ammonium ions instead of the energy in light.

Either as ammonium ions or as nitrate ions, these nitrogen compounds can be absorbed by flowering plants and used in biosynthesis. This completes the basic nitrogen cycle which is summarised in Figure 10.3.

There is more to the nitrogen cycle. Nitrogen compounds are lost from soil because

o ammonium and nitrate ions readily dissolve in water and are continually drained away (this process is called **leaching**) to end up in rivers, lakes and the sea

o certain bacteria, called **denitrifying bacteria**, carry out **denitrification** by using oxygen in nitrate ions in their respiration to leave behind only nitrogen gas (which flowering plants cannot use)

o humans remove plants and animals from the cycle and eat them, but nitrogen compounds from humans' urea and faeces may, instead of being returned to the land, be released into rivers, lakes and the sea

Nitrogen compounds are added to soil because

o electrical discharges in the form of lightning combine the oxygen and nitrogen in air to form ions which dissolve in rain and return to the soil as nitrate ions

o certain bacteria, called **nitrogen-fixing bacteria**, are able to use nitrogen gas from air to make organic nitrogen compounds for their own bodies and even for some leguminous flowering plants (leguminous plants are those forming pods, such as beans and peas)

o humans use inorganic nitrogen fertilisers containing ammonium and nitrate ions

In Figure 10.4 these processes have been added to the basic nitrogen cycle. The complete cycle looks complicated but it makes sense. You will find it easier to understand and remember if you draw the basic cycle and add to it the extra

124   processes that remove and add nitrogen compounds.

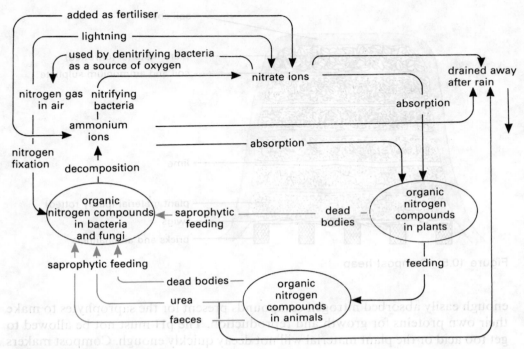

Figure 10.4 Nitrogen cycle

The diagram contains the following labels:

added as fertiliser

lightning

used by denitrifying bacteria
as a source of oxygen

nitrate ions

drained away
after rain

nitrogen gas
in air

nitrifying
bacteria

absorption

ammonium
ions

nitrogen
fixation

absorption

decomposition

organic
nitrogen compounds
in bacteria
and fungi

saprophytic
feeding

dead
bodies

organic
nitrogen
compounds
in plants

saprophytic feeding

dead bodies

urea

faeces

organic
nitrogen
compounds
in animals

feeding

10.5 Compost

Compost is half-decayed plant material which is added to soil as fertiliser. Microorganisms in the soil complete the decay and so provide the ions which the plants need for growth. Because compost is made from plants that have grown successfully, it contains all the elements needed for plant growth, whereas inorganic (artificial) fertilisers may contain only a few elements. Another advantage of compost over inorganic fertilisers is that it improves the drainage of the soil yet allows the soil to hold enough moisture and air for healthy growth of roots. But there is no reason why both compost and inorganic fertilisers should not both be used on soil.

To make compost, vegetable matter from kitchen waste such as potato peelings, old cabbage leaves and bean pods, together with any other soft plant waste from the garden such as grass mowings and fallen leaves, is put in a heap. Woody plant material should not be put in the heap because it takes too long to decay. As in making silage, the saprophytes which carry out the processes of decay must be encouraged by giving them optimum conditions for growth.

A compost heap must contain air to provide oxygen for aerobic respiration by the saprophytes and small animals in the heap. Their aerobic respiration is essential to provide heat to kill weed seeds. A compost heap must not get too dry or the saprophytes will be unable to digest and absorb their food. On the other hand, it must not get too wet or the air in the heap will be reduced. There must be 125

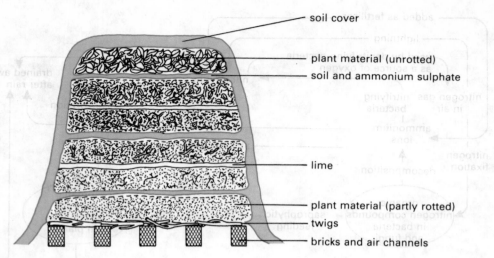

soil cover

plant material (unrotted)

soil and ammonium sulphate

lime

plant material (partly rotted)

twigs

bricks and air channels

Figure 10.5 Compost heap

enough easily absorbed nitrogen compounds present for the saprophytes to make their own proteins for growth and reproduction. The pH must not be allowed to get too acid or the plant material will not decay quickly enough. Compost makers use different methods to provide these conditions. Figure 10.5 shows one method.

Rows of bricks at the bottom of the compost heap provide air channels. Woody twigs laid across the rows stop too much of the compost falling down into the air channels. About 20 cm of plant material are added, followed by a sprinkling of ammonium sulphate, a cheap source of nitrogen, and a thin layer of soil to make sure the right bacteria are present. Another 20 cm of plant material are added, followed by a sprinkling of lime, to prevent the heap becoming too acid. Lime comes in different forms: chalk (calcium carbonate) and hydrated lime (calcium hydroxide) are the most suitable.

The layers of plant material, ammonium sulphate, soil, plant material and lime are repeated twice more, making the heap just over a metre high. A suitable area for the heap is a metre square. Some people enclose the heap with wire netting or wooden planks to hold it in place. A light covering of soil will also help to keep it in place. In dry weather the heap may need watering and in very wet weather it may need covering. Some people drive stakes into the sides of the heap and then remove them to make more air channels.

A compost heap is built up gradually as plant material becomes available: by the time material is added to the top layer, the material at the bottom may already be well rotted. When a compost heap is high enough, a new one is begun. In summer compost is ready after only a few weeks; it takes longer in winter. If the top surface of the heap has not rotted when the material lower down is ready to use, it can be removed and added to the new heap.

A vast number of small animals such as earthworms and wingless insects called springtails feed on the vegetable matter in the heap. These are the animals in the heap that need oxygen for their respiration. The faeces of these consumers contain undigested but small particles of plant material which the saprophytes can decay

126

more easily than, for example, an old cabbage leaf. In a well made compost heap the temperature rises due to the respiration of both saprophytes and consumers: it should rise high enough to kill the seeds of weeds which may have been put on the heap; surviving weed seeds are a nuisance when they germinate after the compost is spread on the soil. The high temperature does not last long and is no danger to animals, such as earthworms, who move away from it. Good compost is dark brown, crumbly with a faint earthy smell.

You can make a compost heap without all this special treatment: left to itself, plant material decomposes anyway, but it takes a lot longer.

Questions

Q 10.1. Look at the water cycle shown in Figure 10.6 and identify the processes A, B, C, D, E and F. Choose answers from the processes given below.

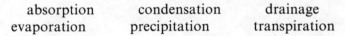

| absorption | condensation | drainage |
| evaporation | precipitation | transpiration |

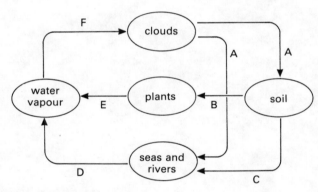

Figure 10.6   Water cycle for Question 10.1

Q 10.2. Look at Figure 10.4 of the nitrogen cycle and explain how the following processes are carried out:

denitrification   leaching   nitrification
nitrogen fixation   decay (putrefaction)

Q 10.3. Every few years a farmer may sow a mixture of grass and clover seeds in a field, instead of the usual crop, and allow cattle to graze on it. Clover is a leguminous plant whose roots contain nitrogen-fixing bacteria. Before sowing his usual crop again, the farmer ploughs the grass and clover into the soil. Suggest why this should increase the yield from the farmer's usual crop.

Q 10.4. Why must a good compost heap contain (a) air, (b) moisture, (c) easily absorbed nitrogen compounds (such as ammonium sulphate), (d) consumers and (e) saprophytes?

127

PLANT CROPS

11.1 Commercial crop production

The three plants, parsley, mint and chive, shown in Figure 9.1 are examples of crops grown commercially to sell to markets and shops for food. The parsley plant was grown from seed, the mint from a separated shoot and the chive from a small bulb separated from a cluster of bulbs. Plants grown from seed (as the parsley was) are the result of sexual reproduction. Plants grown by other methods (as the mint and chive were) are the result of asexual (not sexual) reproduction. In the commercial crop production of flowering plants there are many different methods, but all must use either sexual reproduction or asexual reproduction.

In sexual reproduction, the new nucleus from which an offspring grows (inside a seed) gets DNA from both a male nucleus and a female nucleus. This is why offspring from sexual reproduction are nearly always different from their parents and from one another (see Section 9.2). These differences between offspring are called **variation**.

In asexual reproduction the DNA in offspring comes not only from one parent but from one nucleus or from identical nuclei: the result is that in what they inherit the offspring are exactly like the parent and exactly like one another. The production by humans of genetically identical offspring, with exactly the same DNA, is called **cloning**.

(a) Seeds

Seeds are the result of sexual reproduction. The embryo flowering plant inside a seed has inherited half its DNA from the male nucleus in the pollen tube and the other half from the female nucleus in the egg cell in the ovule. Flowering plants usually produce large numbers of seeds, whereas asexual reproduction usually results in few offspring. Under natural conditions most seeds are 'wasted' because they do not fall to the ground in places where they can grow: if a plant did not produce a large number of seeds, it would not survive in the wild. Commercial growers of certain plants want large numbers and may not be too concerned if there is a slight variation among them: for them, crop production from seed is ideal because they can make sure that the seeds go into suitable soil. Parsley is only one such crop: cereal crops, such as wheat, oats, barley and rice, are grown from seeds on a large scale.

All seeds contain food, which the embryo uses when it begins to grow. This is what makes so many seeds, such as beans, valuable as food for humans: we eat the food stored for the embryo's growth. The embryo is well protected in the seed and can survive, lying **dormant** (inactive), for months, even years, in dry cold conditions. When the right conditions are provided, the embryo begins to grow 129 (germinates). The food stores in the seed are enough for the embryo to grow into a small plant able to make its own food by photosynthesis and able to obtain its own inorganic ions. Growth of the embryo, until it is an independent plant, is called **germination**.

The right conditions for germination are
○ a supply of water (but not too much)
○ the right temperature (which is different for different plant species)
○ sufficient air (to provide oxygen for respiration)

In the UK the climate in Spring out in the open cannot be relied on to provide these conditions: it may suddenly become too cold or too wet. Commercial growers, especially farmers, may have to sow their seeds outside hoping that conditions will stay right throughout the process of germination and that enough seedlings (newly germinated seeds) will survive to grow into a crop. All over the world there is risk when seeds are sown outside: a greater risk with some crops than others; a greater risk in some places than others. Often it is too expensive to do anything else but sow the seeds outside and hope.

Tomatoes have long been grown inside in the UK: the cost of growing them inside is less than the cost of losing a large proportion of the crops grown outside.

(b) Cloning

Cloning is reproduction (brought about by humans) in which nuclei divide to give offspring that all have exactly the same DNA in their nuclei. Such reproduction is **asexual** (not sexual). The offspring with the same DNA are called a **clone**.

Humans began cloning plants centuries ago, long before anyone knew about DNA. Growers, gardeners and farmers have always tried to reproduce successful plants by asexual methods that give offspring with the same characteristics as the parent plant. Cutting small shoots off the parent plant and encouraging these shoots to develop roots and grow into independent plants is a common method. Figure 11.1 shows plants being **propagated** (multiplied) from shoots in this way. Cloning is the only way they can reproduce exactly a plant that they want more of.

(i) Vegetative propagation

When asexual reproduction involves forming offspring from plant organs such as stems, roots, leaves, bulbs or any organ other than a seed, it is described as **vegetative propagation**. The separation and growth of the small shoots in Figure 11.1 is vegetative propagation. Many of the plants that we eat are the result of vegetative propagation: for example, potatoes, yams, ginger and garlic are grown from plant organs. This gives the grower plants that have exactly the same characteristics as the parent plant.

The usual methods of vegetative propagation have disadvantages. Taking stem cuttings and rooting them, for example, is sometimes unsuccessful, may take several months even when it is successful, and is always labour-intensive. To plant the cuttings people have to spend a lot of time preparing the right size of shoot, taking off some of the leaves, making holes in the soil or rooting mixture, putting each shoot in its own hole and pressing the soil or rooting mixture firmly about it. This is what was done with the shoots in Figure 11.1. In Figure 11.2 you can see a *Chrysanthemum* plant that was propagated in this way.

Figure 11.1 Plant propagation

Figure 11.2 *Chrysanthemum* plant

(ii) Micropropagation

A new technique, developed since the 1950s, is still expensive and is used commercially only for valuable plants. It will become cheaper and more widely used. It is a method of cloning called **micropropagation** or **tissue culture**, in which small pieces are cut from a parent plant and grown *in vitro*. The small piece of plant tissue (called an **explant**) is put on a sterile nutrient medium. Even if the explant is only a single cell, it has the DNA of its parent plant.

From the explant there develops a mass of growing and dividing cells called a **callus**. If the nutrient medium contains an **auxin**, a substance that encourages the callus to grow in an organised way, small plant-like shoots develop on the callus. Figure 11.3 shows a callus which has grown from four pieces of hyacinth stem that were put on a nutrient medium containing an auxin. Two small shoots can be seen starting to grow on the callus (the colourless mass of tissue between the four pieces of stem).

Micropropagation has been used to grow orchids, date palms, pineapples and rubber plants for many years. The plants are propagated more quickly than from seed and other plant organs and are disease-free. They are easily transported world-wide as tiny recognisable plants (at a slightly later stage than in Figure 11.3) growing in a nutrient medium in a culture bottle. Micropropagation is used increasingly by plant breeders to multiply stocks of new varieties quickly: new varieties of carnations, cauliflowers, *Alstroemeria* and exotic species.

11.2 Controlled environments

Instead of planting seeds or growing stem cuttings out in the open, where the climate cannot be controlled, growers try to provide certain plants with suitable controlled conditions for growth. The place for this is the **glasshouse** (or greenhouse). Here plants can be grown summer and winter no matter what the temperature, rainfall and light intensity outside.

Figure 11.3   Small shoots developing on a callus by micropropagation

(a) Glasshouses

Many glasshouses are still controlled entirely by people in the old-fashioned way. Plants are grown in soil or compost in pots or troughs; a heater is switched on if it gets cold; the windows are opened if it gets hot; a sprinkler is turned on if it gets dry; fertilisers are added to the watering system when they are needed.

The most up-to-date glasshouses are fully automated: sensors measure the conditions and control systems correct them. The temperature is corrected by ventilation, shading or artificial heating; the humidity (dampness) is kept up by a mister spraying a fine mist into the air; the carbon-dioxide level is kept up by releasing carbon-dioxide gas; light intensity can be increased by artificial lighting or decreased by shading; the fertiliser solution provided for the roots is of different strengths at different stages of growth. Such control makes it possible to grow seasonal flowers at any time of the year: chrysanthemums, for example, are grown in a completely artificial light-dark system.

Automated glasshouse control is made easier when crops are grown in the glasshouse by a system known as **hydroponics** or soil-less water culture. The plants get water and ions from a fertiliser solution. The solution provides different proportions of water and ions at different stages of the plants' growth. Although there is no soil, each plant grows in a sterile supporting medium such as sand, gravel or a manufactured porous sponge-like product. The fertiliser solution is pumped along a pipe between the rows of plants: young plants are given about seven fertiliser feeds a day. Figure 11.4 shows young tomato plants grown hydroponically.

132

young tomato plant

sterile supporting medium

fertiliser solution

Figure 11.4   Tomato plants grown hydroponically

In between the manually controlled glasshouse and the fully automated one are the majority of the systems which rely on people and machines. Look at the example in Figure 11.1. A humidity sensor has been stuck in the ground near the cuttings: if the humidity gets too low, the mister will be switched on automatically. But there is nothing automatic about the glass at the side of the glasshouse: someone has whitewashed it to reduce direct sunlight which would overheat the cuttings.

Glasshouse products cannot usually be sold at competitive prices when products grown outdoors in the same country are available fresh. (Tomatoes are one of the exceptions.) All glasshouses are expensive: there is the cost of the glasshouse itself and of the equipment, the cost of repairs and the cost of maintaining the necessary conditions in the glasshouse, whatever the climate outside. Out-of-season glasshouse-crop production can be profitable for the grower only if the crop can be sold at a high price which covers all these costs. This will not be possible if it is cheaper to import the crop from a country where it is grown out of doors.

Because of the high cost of fuel, glasshouses growing tomatoes throughout the year are less profitable on the colder eastern side of the UK than in the warmer south west. Glasshouse production of expensive flowers is profitable all over the UK.

(b) Cloches
Cloches or frames are movable tent-like or tunnel-like covers of glass or transparent polythene which many commercial growers put over soil and plants in the open. They do not control growing conditions but merely make them less extreme. The cloche shown in Figure 11.5 keeps off frost and extremely wet conditions in winter, gives the covered plants some protection from wind, and helps to warm the soil in Spring.

Cloches do not need to be lifted for plants to be watered because rain water seeps sideways as well as draining downwards through the soil, while the plants'   133

Figure 11.5   Lettuces under a tent cloche

roots grow outwards and downwards towards the water. Crops such as strawberries and lettuces are ready to send to market much earlier than they would be without the protection of cloches. Seed germination is far more successful under a cloche than in the open.

Cloches have advantages over glasshouses in being much cheaper and needing far fewer people to look after them. They can easily be moved about. The soil below them is covered only for part of the year and gets time to recover when it is open to the weather. In a glasshouse, the soil or compost in which plants are grown must be regularly replaced.

(c) An automated hydroponic unit

At the other extreme from the simple cloche are complicated systems such as the Landsaver unit. Figure 11.6 shows a large windowless box-like insulated unit in

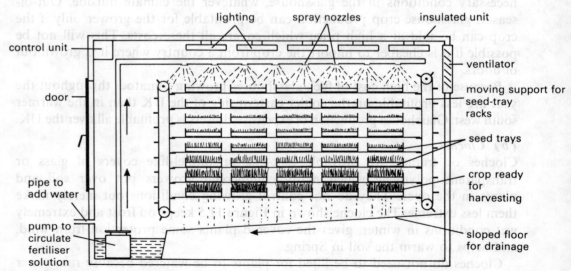

Figure 11.6   An automated hydroponic unit

which grass is grown hydroponically (in water without soil) under rigidly controlled conditions. Landsaver is the company that developed it.

Grass grown from seeds in trays in a Landsaver unit is ready in eight days to be fed to cattle and other animals such as goats and horses. The internal temperature is kept at 20°C. Overhead fluorescent lamps provide continuous lighting. The unit is connected to a mains water supply but uses only 2 per cent of the water used by field-grown grass even in dry areas. Fertiliser, which provides all the inorganic ions needed, is added to the water, which is sprayed on the trays for three minutes every six hours.

A Landsaver unit of 10.8 m × 2.9 m × 3.5 m high can supply 1 tonne of fresh green grass every day of the year regardless of the weather. Fresh seed is sown daily in trays placed on racks at the top of the unit. Each day the trays of growing grass are moved downwards. Trays reach the base of the unit after eight days, by which time they contain a mass of tightly packed rooted grass which can be fed, whole or chopped, to the animals. Trays are washed in disinfectant.

Buying a Landsaver unit and having it connected to mains water, mains electricity and a suitable drainage system means investing a lot of money. More money has to be spent on its running costs, which include labour, seed, electricity, water, fertiliser and disinfectant. The advantages are a continuous supply of fresh disease-free grass and the freeing of land, which would otherwise be used to grow animal food, for the growth of other crops.

11.3 Competition

Every plant needs a certain space in which to grow. From its surroundings a plant must get water, inorganic ions, carbon dioxide and light. Plants that are growing close together **compete** with one another for anything that is in limited supply, such as water, inorganic ions and light. In glasshouses they may compete for carbon dioxide. In the open they do not compete for carbon dioxide because there is no shortage of it: as plants use carbon dioxide, it is replaced by diffusion from the surrounding air.

If the competing plants are of the same species, as when a single crop is planted too closely, it is **intraspecific competition**; if they are of different species, as when weeds compete with a crop, it is **interspecific competition**. If a plant cannot get enough water, inorganic ions or light, all of which it needs to make more cells and tissues, it will grow slowly or not at all; eventually it may die. When all the plants in a crop are competing with one another, all of them will be affected to some extent.

(a) Intraspecific competition
Gardening books and seed packets give precise distances at which plants should be grown from one another. Usually planting is done in rows: for French beans, for example, the gardener is advised to allow

15 cm between seeds and 60 cm between rows

Figure 11.7 shows that plants sown in rows may compete with one another all   135

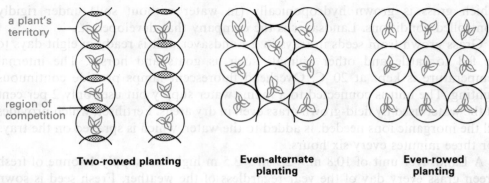

a plant's territory

region of competition

Two-rowed planting   **Even-alternate planting**   **Even-rowed planting**

Figure 11.7   Planting patterns

the way along a row but not between the rows. It seems that space between the rows is wasted. But there may be other reasons why such spacing between the rows is suggested: for example, it makes it easier to hoe away the weeds and to walk between the rows to pick the beans.

For the commercial grower these reasons no longer apply: the commercial grower removes weeds not by hoeing but by spraying them with weed-killer (herbicide); beans are harvested not by hand but by machine.

Look at the even-rowed planting and the even-alternate planting in Figure 11.7. In both, each plant's territory touches but does not overlap with its neighbours' territories. Both waste much less space than planting in separate rows, but you can see that even-alternate planting wastes the least space of all.

Giving each plant a large enough territory to grow to its maximum size may not be the best commercial policy. In the past even commercial growers gave plants too much space; closer planting may result in smaller individual plants but a higher total yield. For example, French beans give a maximum yield with 3–4 plants in an area 30 cm × 30 cm with an even-alternate pattern of planting even though the plants would be bigger if grown less close together.

(b) Interspecific competition

If weeds are allowed to grow between crop plants, they take up some of the water and inorganic ions and their leaves shade the crop plants from the light. These are the same effects as when there is competition from plants of the same species. Because weeds tend to grow quickly and vigorously, the effects are usually more serious. Crop plants with quick-growing large leaves are better able to compete with weeds than others: their leaves shade the weeds from light and weed growth, rather than crop growth, is slowed. The even-alternate pattern of planting reduces the spaces where weeds can grow.

Onions have long slender leaves which are less able to shade the weeds and therefore do not compete well. An experiment was carried out to see how weeds affect an onion crop. Weed seeds were scattered among three-week-old onion seedlings on five identical weed-free plots. A sixth identical weed-free plot of onion seedlings had no weed seeds scattered on it and was kept weed-free throughout the experiment. At different times each of the other five plots had all the weeds

136

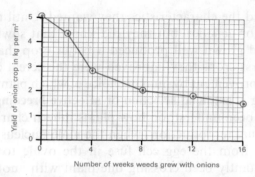

Figure 11.8 Effect of weeds on an onion crop

removed from it and after that was kept weed-free. All the onion seeds were sown in May and all the onions harvested in October.

The graph in Figure 11.8 shows the effect on the yield of the onion crop of allowing weeds to grow among the three-week-old onions for different periods of time. Where no weeds were allowed to grow at all, the crop yield was 5 kg per m²; where weeds were present for most of the growing period, the crop yield was only 1.5 kg per m².

11.4 Plant breeding

When carrots are planted too close to one another, some carrots survive and grow quite well, others survive but do not grow well and others do not survive at all. When the carrots are dug up, even though all have suffered from competition, the difference in size between the largest and smallest is obvious. In other words, some carrots are better at competing than others: they may grow quicker; they may have longer and fatter roots, bigger leaves or more efficient chlorophyll; or they may have other advantages.

Left alone, the carrots that grow better will produce more seeds than the poor competitors. If all the seeds fall to the ground and grow the following year, there will again be competition between the different plants and again only the best competitors will survive to produce seeds for the following year. This happens without interference from humans: it is an example of **natural selection**. Carrot plants with the DNA that gives them an advantage over other carrot plants will survive to pass that DNA on to seeds and to offspring grown from them. In the next generation, therefore, there will be more carrot plants with what in the previous generation was an advantage. In this way the average quality of the carrots can improve with each generation, all without interference from humans. This is a simple example of **evolution** or change over the generations. (It is by natural selection over millions of years that apes and humans have evolved from a common ancestor.)

But human interference will make it possible to improve the quality of carrots far more quickly. For thousands of years humans have deliberately grown crops 137 from plants that had the characteristics they wanted. Until recently they did not know about DNA or genetics, but they did know that sowing seed from plants that themselves had grown well produced a better crop than sowing seed from poor plants. The first crops they improved were the cereal crops such as wheat, oats and barley. Improvement by human choice of which plants (or which animals) shall have offspring is **artificial selection** or **selective breeding**.

Ever since it has been understood how plants inherit characteristics from both their parents in sexual reproduction (when the male nucleus from the pollen and the female nucleus from the egg cell fuse in the ovule to form a seed), plant breeders have constantly been crossing one plant with another to try to obtain the best characteristics of both parents. They do this by transferring pollen from the anther of one plant to the stigma of another. Even this is a hit-or-miss affair because only half the DNA of each parent is passed on, and breeders cannot be sure that what they want is passed on.

Kiwifruits grow wild in China. In 1904 a farmer in New Zealand sowed a batch of kiwifruit seeds from China and in 1910 they bore fruit. Although there was mild commercial interest in the fruit, it was not till 1960 that another farmer, Hayward Wright, planted a large number of different seeds from the descendants of the original batch in order to find plants with the best characteristics. One plant, now called the Hayward variety, was much better than the rest: most of the world's commercially grown kiwifruits are now produced by vegetative propagation from this variety.

This story makes two points: plants with different characteristics grew from the different seeds as a result of sexual reproduction, but, in order to get more of the variety with the characteristics that he wanted, the farmer had to reproduce the plants asexually.

Although plant breeders continually try to develop varieties that give a higher yield than existing varieties, there are other important characteristics that they look for. Crops harvested by machines need to grow at the same rate and ripen at the same time. Because plants in crops harvested by machine are grown closer to one another, they also need to be especially disease-resistant. Last but not least, crop plants must look attractive to customers if they are to sell well. The plant breeder tries to get all these characteristics (and others as well) in a single variety.

The attempt is still made mainly by pollinating a plant that has many good characteristics with another that has many good characteristics in the hope that a plant which has most of the good characteristics of both parents will grow from one of the seeds. One day genetic engineering (see Section 25.4) should be able to isolate and transfer exactly the right pieces of DNA from one plant to another and so produce offspring with *all* the good characteristics.

Questions

Q 11.1. Explain why a glasshouse grower would want to increase the
138 (a) amount of carbon dioxide in the air

(b) light intensity
(c) air humidity

Q 11.2. A journalist wrote the following about micropropagation:

The day may come when allotments and glasshouses will be planted with clones as well as seeds, bulbs and cuttings.

What is wrong with this statement?

Q 11.3. Look at the graph in Figure 11.8.
(a) What would have been the onion yield if weeds had been removed after six weeks?
(b) In which four-week period did the weeds have the greatest effect on onion yield?
(c) Describe and explain the trend in onion yield as weeds were allowed to grow with the onions for longer periods of time.

Q 11.4. Many plants, such as potatoes and strawberries, are grown out of doors in soil that is covered by a sheet of black polythene. Shoots of the plants grow into the air through slits in the polythene. Suggest reasons why covering the soil with black polythene increases the yield of the crop.

CROPS AS ECOSYSTEMS

12.1 Introduction

An **ecosystem** is made up of a community of many different living organisms and their non-living **environment**, which includes climate and soil. Natural ecosystems are places such as oak woods, ponds, deserts and the seashore. They are easily recognisable by the plants and animals in them as well as by their non-living parts such as fresh flowing water in a river ecosystem or dry sand in a desert.

When farmers grow crops over large areas of land, they make new ecosystems. A wheat field, a rice field and a field of cabbages are all ecosystems easily recognised by both the plants and animals in them. For example, you would find cabbages and cabbage-white butterflies or the caterpillars of cabbage-white butterflies in a cabbage field but not in a field of wheat.

In natural ecosystems, elements such as carbon and nitrogen are recycled (see Sections 10.3 and 10.4) and so is water (see Section 10.2). In crop ecosystems many elements are not recycled: they are taken out of the ecosystem when crops are harvested and sent away as food for humans. The farmer needs to return the elements that have been taken away by adding fertilisers, usually inorganic ones, to the soil. Water is usually recycled naturally: but, if there is not enough rain, the farmer waters the crop or adds water by irrigation of the land.

All ecosystems contain **producers** (plants) and **consumers** (animals and most microorganisms) (see Section 8.1). **Decomposers** (saprophytic bacteria and fungi) are special kinds of consumers that feed on dead organic material which they digest externally. Consumers can be further subdivided according to what they eat: animals that eat plants are **herbivores** or **primary consumers**; animals that eat animals are **carnivores**. If an animal eats animals that eat plants, it is a **secondary consumer**; if an animal eats animals that eat animals that eat plants, it is a **tertiary consumer**; and so on.

For example, the caterpillar that eats a cabbage leaf is a primary consumer; the robin that eats the caterpillar that eats the cabbage leaf is a secondary consumer; the cat that eats the robin that eats the caterpillar that eats the cabbage leaf is a tertiary consumer; the fox that eats the cat that eats the robin that eats the caterpillar that eats the cabbage leaf is a quaternary consumer. This is called a **food chain**. It can be shown as

cabbage	→	cabbage-white butterfly	→	robin	→	cat	→	fox
producer	→	primary consumer	→	secondary consumer	→	tertiary consumer	→	quaternary consumer

141

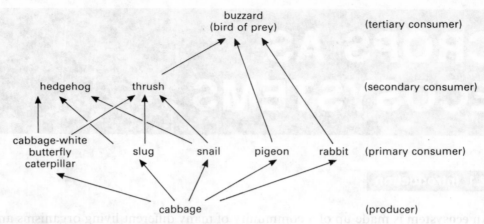

Figure 12.1 Food web in a cabbage field

The food chain is a simplified feeding relationship. In a natural ecosystem there will be a large number of different plants at the producer level and a large number of different animals at the different consumer levels. Even in a field of cabbages kept free of other producers (weeds), caterpillars will not be the only primary consumers: slugs, snails, pigeons and rabbits, for example, also eat cabbages. At the secondary-consumer level, birds of prey, such as buzzards, eat pigeons and rabbits; both birds, such as thrushes, and hedgehogs eat cabbage-white butterflies, slugs and snails. Buzzards also eat thrushes, when they become tertiary consumers.

All these feeding relationships are described as a **food web**. Figure 12.1 shows a simple food web in a cabbage field. An animal that eats at different levels, such as the buzzard, is described as belonging to the higher or highest of these levels.

12.2 Plant–crop ecosystems

In the ecosystem of a cabbage field, the cabbages are the producers. The farmer who has planted this crop wants to sell it for humans to eat: the farmer wants humans to be the primary consumers. The farmer will do everything possible to prevent other primary consumers, such as cabbage-white-butterfly caterpillars, from eating it. Any organism that damages a plant crop or reduces its yield is a **pest**. Primary consumers that eat the crop before it is harvested are pests; so are fungi that cause diseases of the crop; so are the weeds that compete with the crop for light, water and inorganic ions. Chemicals that kill pests are called **pesticides**.

Figures 12.2 and 12.3 show two primary consumers that farmers try to keep away from their crops: a millipede and a slug. Both these animals are pests that eat plant parts: millipedes are found in the soil eating underground parts of plants such as potatoes and carrots; slugs usually feed above ground on leaves of plants such as cabbages and lettuces.

Other examples of primary consumers that farmers try to keep away from their crops are nematodes (very small eelworms), which eat potatoes, and aphids (small

142

Figure 12.2   Millipede

Figure 12.3   Slug

blackfly or greenfly), which suck the sugary sap from the shoots of beans and fruit trees. Aphids not only take from plants food that should be going to the seeds and fruits; as they fly about they carry pathogens (disease-causing microorganisms) from infected to healthy plants.

Organisms that spread pathogens are called **vectors**. Vectors are pests because, by spreading pathogens, they reduce the yield of the crop. Aphids are pests both because they are primary consumers and because they are vectors. Different pesticides exist which will kill pests. One problem for the farmer is how to get the right pesticide within reach of the pest. Another problem is to find a pesticide that does not kill useful organisms as well as pests.

You may think of millipedes and centipedes as similar. Compare the millipede in Figure 12.2 with the centipede in Figure 12.4. The difference between them has

Figure 12.4   Centipede

nothing to do with the total number of legs: it is that millipedes have two pairs of legs per segment and centipedes have only one pair. But there is a more important difference between them: while a millipede is a primary consumer eating cabbages and other crops, a centipede is a secondary consumer eating primary consumers: it eats millipedes, nematodes and small slugs, but not large ones. If they know this, gardeners digging over the ground remove millipedes but not centipedes.

This gives a clue to a way of killing pests without using pesticides. Pests (such as slugs, aphids, nematodes and millipedes) are all primary consumers and are all eaten by secondary consumers. Can farmers use secondary consumers to kill primary consumers? It is difficult for farmers to increase the numbers of large mobile secondary consumers such as birds and hedgehogs in their fields. But the time may come when they breed (or buy) centipedes.

The use of living organisms to get rid of pests is called **biological control**. It is an example of biotechnology and can be used instead of chemical control by pesticides. Even biological control has its dangers because it too can have an effect on other organisms and because the effects are difficult to predict. But it is already being used successfully.

In the United States the two-spot ladybird, which is common in the UK, is successfully bred and released in orange groves to eat the aphids (small flies) on the orange trees. Figure 12.5 shows a ladybird and a number of aphids (blackfly) of different sizes, including one of the rare forms with wings.

Biological control also makes use of wasps that lay their eggs in the bodies of young whitefly (one egg per body). The wasps are good at finding young whitefly among the dense leaves of plants. A young whitefly is killed by the development inside it of a wasp from an egg. The wasps do not get rid of all whitefly, but nor do pesticides.

Figure 12.6 shows how biological control by the wasps works. When there are many whitefly, the grower releases the wasps into the glasshouse. Because there are many whitefly for them to eat, the wasps thrive and their number increases. When there are more wasps than whitefly, the wasps are short of food and their

144 Figure 12.5 Ladybird eating aphids

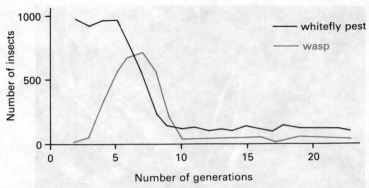

Figure 12.6   Relation between whitefly (primary consumer) and wasp (secondary consumer) in a glasshouse

number decreases. Eventually there is a stable low number of both whitefly and wasps.

Attempts by scientists to control potato eelworm with a fungus that feeds on eelworms (and other nematodes) have not been successful. But we may expect to see more successful uses of biological control in the years ahead.

Living organisms may help farmers and growers in other ways. Fruit growers encourage bee keepers to bring their hives to the orchards when the flowers open. Bees visit flowers to collect **nectar** and pollen for food. (Nectar is a sweet sugary secretion from some flowers.) By searching for nectar and pollen in the flowers, bees transfer some of the pollen from anthers to stigmas: they carry out pollination of the flowers and increase the number of fruits that form after successful pollination.

Figure 12.7 shows a bee on a raspberry flower with its pollen basket on its hind leg full of pollen. Some of the pollen is sure to have been brushed from the

Figure 12.7   Bee on a raspberry flower

145

Figure 12.8 Earthworm

anthers on to its body and then brushed from its body on to the stigmas of the flowers.

Another animal that helps the farmer is the earthworm. Earthworms make burrows in the soil to live in. These burrows help drainage of water from the soil and increase the air supply to roots. There is yet another important way in which they improve the soil. An earthworm eats soil containing dead organic material such as leaves. Its food passes through its body: what cannot be digested and absorbed is expelled as faeces in the form of a **worm cast**. The soil in a worm cast is fine and crumbly and improves the texture of the soil. It contains undigested material that is broken up in a form that microorganisms can easily decompose: earthworms therefore help in recycling elements.

12.3 Chemicals for the grower

Every year the research departments of large chemical firms test thousands of different chemicals for their effectiveness in killing pests. Many pesticides are effective yet have disadvantages. They may be dangerous poisonous substances which must be handled by humans with great care. Yet it is impossible to be sure that the instructions printed on them are always followed: in foreign countries the language in which the instructions are printed may not be understood; or the container may get dirty and the instructions be impossible to read.

A recent medical study at Imperial College, London, has suggested that tens of thousands of farm workers die every year around the world from poisoning by pesticides. Some of the dangers of pesticides may not be obvious until they have been in use for several years.

(a) Insecticides

Insecticides are chemicals that kill insects. They have had some of the most
146 damaging and unexpected effects. Chemicals that kill insect pests also kill other

insects. They may kill bees which are valuable in pollinating flowers, including those of fruit crops. They may kill ladybirds, the valuable secondary consumers that eat aphids. In killing pollinators and secondary consumers, insecticides may even do more damage than that done by the original pests.

Secondary consumers such as birds and mammals are not killed directly by an insecticide, but they have less to eat once the insects have gone. Tertiary and quaternary consumers also have less to eat when there are fewer secondary consumers. The balance of the ecosystem is upset. When the number of secondary consumers is reduced, other pests which they were keeping in check will increase in number and may cause more damage than the original pests.

Some years ago there was a problem with an insecticide called DDT. DDT was good at killing insects: large quantities of it were used, especially to kill mosquitoes in countries where mosquitoes were the vectors of the microorganisms that caused malaria. At the time it was not realised that DDT was a very long-lasting chemical. It stayed in the soil where it was sprayed for up to ten years: plants grown in the soil absorbed the DDT. When humans ate these crops, the DDT was not excreted but remained in their lipid (fat) tissues. Table 12.1 shows average DDT levels in human lipid in four different countries in the 1970s.

No known harm was caused by the DDT to the humans who absorbed it, but that was good luck. For all we knew beforehand, the effects might have been serious. DDT did in fact harm animals. DDT appeared in the lipid of animals throughout all the food webs where primary consumers fed on sprayed crops and where DDT drained from crops into waterways. It was found even in the penguins of the Antarctic: presumably it had been in fish they had eaten.

Because DDT is so soluble in lipid, it builds up in the lipid of each animal as it passes from animal to animal in the food chain. Look at Table 12.2, which shows how DDE (a breakdown product of DDT) was concentrated in the lipid of organisms in a food chain. Although there were only 3 parts per million in the fallen leaves, it was not excreted from the earthworms. The earthworms ate many leaves and the DDE built up in their lipid to 33 parts per million. In the robins who ate the earthworms the DDE built up to 875 parts per million.

Harm was caused to the animals even further along the food chain: birds of prey that ate the robins had even more DDE in their bodies. They did not breed

Table 12.1 DDT in human lipid

Country	DDT in lipid in parts per million
UK	3
USA	6
Israel	19
India	31

Table 12.2 DDE in a food chain

	Concentration of DDE in parts per million
Fallen leaves	3
Earthworms	33
Robin	875

successfully because the shells around their eggs were abnormally thin and broke easily. Some birds of prey, such as peregrine falcons, became very rare in the UK.

(b) Fungicides

Fungicides are chemicals that kill fungi. Fungi living as parasites (see Section 7.4) or saprophytes (see Section 7.2) damage seedlings, growing crops and stored food. Most fungicides prevent infection from spores or hyphae only on the surface of a plant. These fungicides do not get into the plant, do not protect new growth and are easily washed off by rain. They have to be sprayed on the crops several times in a growing season.

In recent years fungicides, such as benomyl, have been developed which are absorbed by the plant and kill fungi inside it. An ideal fungicide has not yet been developed. The effect of fungicides on food chains is not well understood.

(c) Herbicides

Herbicides are chemicals that kill plants. Growers want to kill weeds, the pests that compete with the crop for valuable water, inorganic ions and light (see Section 11.3). It is not difficult to find a chemical that kills plants: what is difficult is to find a chemical that kills weeds but not crop plants. Such herbicides do exist for certain crops. They are called **selective herbicides**. The chemicals that kill weeds but not grass in a lawn are selective herbicides.

Scientists are hoping to use genetic engineering (see Section 25.4) to develop crop plants which will produce an enzyme that will destroy a herbicide. The herbicide could then be sprayed on both crops and weeds, but the crops' enzymes would make the herbicide on their leaves harmless. This would be a mixture of chemical technology and biotechnology.

Herbicides have had a great effect on ecosystems throughout the world. Plants that we call weeds are the food of many primary consumers. There are many instances where not only have the weeds disappeared but so have primary consumers, such as butterflies, which depend on them.

(d) The disadvantages of pesticides

Whenever a pest is killed, there are effects on the food chain. The consumers that fed on the pest have less food and they decrease in number. If they were also eating other pests, those pests may now increase in number. When insecticides were used to kill insect pests of fruit, the secondary consumers that fed on the pests were reduced in number. These consumers also ate the red-spider mite. The result is that the red-spider mite has now become a major pest, and it is one resistant to many pesticides.

We have seen that, when a pest is killed, its consumers decrease in number. There may not be enough of them to keep the pest under control if and when it again begins to multiply. Because its natural enemies have disappeared, farmers may end up with greater numbers of the pest than when the pesticide was first used.

When a pesticide is used a great deal, the pest may develop a resistance to it. This is what has happened with DDT and mosquitoes. It is an example of natural selection: mosquitoes that were not affected by DDT were the ones that survived and they passed on their resistance to their offspring. Where DDT was used a great deal, most of the mosquitoes are now resistant to it.

148

Pesticides are powerful chemicals. They get washed into the soil and spread throughout the waterways. This chemical pollution (see Section 13.3) may kill many organisms some distance from the crop. It took years to discover the harmful effects of DDT. There is always the danger that harmful effects of other pesticides have not yet been discovered.

Biological control is an alternative to pesticides which is cheap, efficient and non-polluting (see Section 13.3). Although it is much safer than pesticides, it too has an effect on natural food chains and so far there are only a few known examples of its successful use. It is unsuitable for use on stored food: no one wants the excretory products and faeces of the consumers mixed with their food.

12.4 Energy in the ecosystem

Plants, the producers, capture some of the energy in sunlight by the process of photosynthesis (see Section 8.2). Although in the UK they capture only about 1 per cent of the light energy that falls on them, they are able to change this light energy into chemical energy in a glucose molecule. This is the most important way in which energy gets into an ecosystem.

About 20 per cent of the energy captured by photosynthesis in the glucose molecule is lost as heat during the plant's respiration. But the rest of the energy (about 80 per cent) remains trapped as chemical energy in glucose and in the many different organic compounds that the plant makes from the glucose as it grows. The energy trapped in this way by three different plant crops (wheat, potatoes and cabbages) grown on a hectare of land is shown in Table 12.3.

When a primary consumer eats a producer (a plant), it takes in as food the chemical energy that the producer has trapped in its organic compounds. But the primary consumer now loses most of that energy. Plants contain a lot of indigestible fibre formed from their cell walls. About 50 per cent of the chemical energy in what the primary consumer eats passes undigested through the gut and is lost as faeces. After digested food has been absorbed through the gut into the body, another 40 per cent of the original chemical energy is either lost as heat

Table 12.3 Energy in food from 1 hectare of land

Food	Energy in megajoules (MJ) per hectare
Wheat	60 000
Potatoes	110 000
Cabbages	20 000
Cattle meat (beef)	5 000
Sheep meat (lamb)	3 000
Pig meat (pork)	9 000
Chicken meat	8 000

149

during respiration or used as energy for movement. Only about 10 per cent of the chemical energy in the original organic compounds of the plant is trapped in the organic compounds of the animal's body. These are approximate figures for UK land mammals. The energy loss can be different in different mammals and in different climates and ecosystems.

The energy trapped in the meat of four different primary consumers (cattle, sheep, pigs and chickens) reared on crops grown on a hectare of land is also shown in Table 12.3.

Table 12.3 shows that one hectare of land can provide humans with about ten times as much food in the form of plants as in the form of meat.

Figure 12.9 shows approximate energy losses after a plant has trapped sunlight and after an animal has eaten a plant.

About the same loss (90 per cent) occurs when a secondary consumer eats a primary consumer and again (90 per cent) when a tertiary consumer eats a secondary consumer and so on. In other words, an animal can use for its own growth only about 10 per cent of the energy in an organism that it eats. The rest (90 per cent) is not digested, is lost as heat or is used by the eater for movement and other activities.

(a) Biomass

The total mass of a number of organisms is called as their **biomass**. Much of it is water but the organic compounds of which the organisms are made contain energy. Many North American and European countries, including the UK, produce far more cereal grains than their populations can eat. With continual improvements in agricultural techniques, these surpluses will increase. Although these surpluses should in the short term be given to countries where there are famines, in the long term these countries want to grow their own food.

Schemes have started in the UK and elsewhere to convert the energy in biomass into ethanol and petrochemicals which can replace the petrochemicals at present obtained from natural gas and oil.

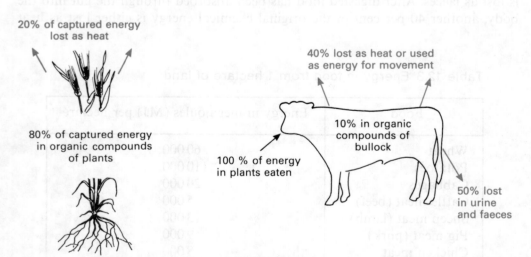

20% of captured energy lost as heat

80% of captured energy in organic compounds of plants

40% lost as heat or used as energy for movement

10% in organic compounds of bullock

100 % of energy in plants eaten

50% lost in urine and faeces

150   Figure 12.9   Energy losses from a plant and from a primary consumer

Q 12.1. Give an example of a primary consumer and a secondary consumer which a farmer might find in a field of wheat. Why might the farmer want to get rid of the primary consumer but keep the secondary consumer?

Q 12.2. Look at the figures in Table 12.3.
(a) Calculate the average energy per hectare trapped in the plant crops. (Appendix A tells you how to calculate an average.)
(b) Calculate the average energy per hectare trapped in the meats.
(c) Explain the difference between the two figures that you calculated in (a) and (b).

Q 12.3. Australian sheep farmers tried planting the prickly-pear cactus from South America as a hedge to stop their sheep from straying. But the cactus spread over their land and was difficult to check. Scientists collected millions of eggs of a South American moth that hatched into caterpillars that ate the prickly-pear cactus. The eggs were scattered over the area in Australia where the cactus plants were growing unchecked. The spread of the cactus was stopped.
(a) Explain why this is an example of biological control.
(b) Why do you think the cactus did not spread over the land in South America?

HUMAN EFFECTS ON THE ENVIRONMENT

More than a million years ago the earliest humans moved through the land in search of food. They killed animals for meat and collected and ate parts of wild plants. Once humans had discovered fire, they stayed longer in one place and began to change their surroundings by clearing away trees and other plants from the places where they cooked, ate and slept.

By 30 000 years ago humans had spread to most parts of the world. They lived mainly by hunting large animals: humans lived together in groups, hunted together and built shelters. By 7000 years ago humans in different parts of the world were growing crops of wheat and oats and keeping domestic animals such as sheep, cattle and pigs.

Once humans started to grow crops they settled in permanent homes and changed their surroundings even more. They cleared wild trees and other plants in order to sow their crops (the start of agriculture) and build their homes (the start of urbanisation).

13.2 Conservation

Few natural environments are left in the world today, almost none in the UK. Nowhere in the UK are there places where plants and animals live completely unaffected by human activities. Many species of plants and animals totally disappeared from the UK as **agriculture** and **urbanisation** (the building of towns) destroyed there **habitats** (the places where they lived). By **conservation** we are trying to keep as many natural habitats and as many different species of plants and animals as we can.

Between 1947 and 1987 the countryside in the UK lost
○ 40 per cent of its ancient broad-leaved woodlands
○ 20 per cent of its hedges
○ 25 per cent of its heaths, moors and wetlands

Of the UK's 55 breeding species of butterfly, one species has become extinct, ten are in danger of becoming extinct and thirteen others are declining; 35 species of birds are also in decline.

The habitats destroyed in the UK were natural habitats that contained a variety of wildlife but they were not the original habitats that covered the land 153 before humans interfered with them. In Africa, Asia and North America too, few of the original habitats remain. But the tropical rain forests in South America are the original habitats which, until the last few years, had not been damaged by humans.

The tropical rain forests of South America are the richest source of life on earth. More than half the world's natural wild animals and plants live there. Thousands of species are still unidentified.

We need to keep these wild habitats because, in the past, they have provided us with

○ 40 per cent of our modern medicines (modern surgery still relies on curare, a South American arrow poison, which relaxes the muscles)
○ food such as coffee, tea, sugar and nuts
○ other products such as rubber, gums and dyes

Although humans now grow all these products as crops, the wild species are a reserve of DNA which can provide many different inherited characteristics which we do not have among the few varieties that we grow. They are a **gene reserve**. Genes are sections of DNA that control what is inherited (see Section 25.3).

When potato blight wiped out nearly the whole potato crop in Ireland in the late 1840s, plant breeders used the wild potato species from the Andes in South America to breed a resistance to this disease into the European potatoes. Recently a wild 'hairy potato' has been used to breed a potato variety which traps aphids among its sticky hairs.

Since the beginning of this century, half the world's tropical rain forests have been cut down to provide hardwood for making furniture in the rich countries and to provide land for crops. Although there is so much life in these forests, the soil is thin above solid rock. Heavy rain may carry away the soil once it is no longer protected by the covering of leaves and held together by the roots of trees.

The soil of the tropical rain forests is in fact unsuitable for growing crops: there are few inorganic ions free in the soil because most of the elements are locked up in the trees (and the other plants of the forest). When the old plants die naturally, the elements are quickly recycled in the warm humid climate to be absorbed by the new plants that grow there. Once the ions have gone, the forests cannot return. The land is then used as rough pasture with cattle grazing at a density of one animal per two hectares (five acres) compared with five animals per two hectares in the UK. Every second 4000 square metres of tropical rain forest are cut down to grow (or try to grow) crops on the land; the trees are often burned to get rid of them. A forest area the size of the UK is destroyed every two years. Conservationists are campaigning to stop this.

(a) Recycling paper

In 1987 131 million trees were cut down to cater for the UK's demand for paper and board. The average family of four threw away six trees' worth of paper and board. Making paper and board from wood pulp is a process that uses a lot of energy: we get most of our virgin paper (paper made from previously unused wood pulp) from countries such as Sweden, who have a cheap energy source in hydroelectricity (electricity from water power).

There is an energy saving of nearly 50 per cent if waste pulp (from paper and board that has already been used) instead of virgin pulp is used. **Recycled paper** is paper or board that contains about 90 per cent old pulp. It used to be of a far lower quality than virgin paper and taking the ink out of it caused pollution. The quality is now excellent: it is no longer a drab grey and can now compete with the highest quality virgin paper. The pollution problems have been largely overcome. At present it is as expensive to make recycled paper in the UK, which has no cheap energy source, as it is to buy virgin paper from abroad. But recycled paper would be much cheaper if there was a greater demand for it in the UK.

In 1987 about 30 per cent of the paper and board used in the UK was collected and recycled. The figure could be as high as 87 per cent. Poor-quality processed waste paper could be used to make animal bedding, house insulation and fuel logs and pellets. The price paid for waste paper has varied over the years; this has made it difficult for groups to organise collection schemes.

A *Recycled Paper Directory*, which includes samples of recycled paper and names and addresses of firms that supply it, is available from Friends of the Earth Trust Ltd, 26–28 Underwood Street, London N1 7JQ (01–490 1555).

13.3 Pollution

Pollution is anything that spoils the environment: it is usually caused by human activities but occasionally by something non-human such as a volcano. Pollution has increased with an increasing population all over the world and with the manufacture of substances that cannot be decomposed by bacteria. It has become difficult to find places where we can get rid of rubbish such as old cars, cans and plastic. Even sewage, which bacteria can recycle, becomes a **pollutant** if too much of it is released in one place at one time.

(a) Sulphur dioxide

The gas **sulphur dioxide** is a serious pollutant of air in Europe, where power stations that burn fossil fuels, such as coal and oil, release up to 85 per cent of the sulphur dioxide found in the air. Sulphur dioxide reacts with the moisture in air to form sulphate ions and then sulphuric acid. With oxides of nitrogen, also given off in the smoke from power stations as well as in car and lorry exhaust fumes, sulphuric acid forms **acid rain**.

There are now strict controls on the release of sulphur dioxide from chimneys in many, but not all, industrialised countries. In Europe, because most winds come from the west, the most heavily polluted areas are to the east of the major industrial cities. There is not much acid rain on the west coast of Ireland or Spain, but there is a great deal in central and north-east England and in central and eastern Europe.

'Pure' rain has a pH of about 5.5; the London smog in 1952 (see Figure 18.8) is now believed to have had a pH of 1.8; lemon juice has a pH of 2.0. Figure 13.1 shows how the pH of a loch in Scotland has fallen since the Industrial Revolution from about 6.0 to about 4.3. Freshwater fish die where there is acid rain, not because of the acid in the rivers and lakes, but because acid releases aluminium 155

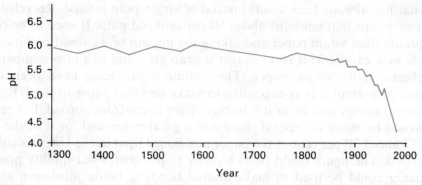

Figure 13.1   Acidity of a Scottish loch

ions in the soil and these drain into rivers and lakes. Aluminium ions cause fish's gills to clog with mucus.

Acid rain has damaged trees by blocking stomata on leaves and interfering with the chemical processes in photosynthesis. The damaged trees become less resistant to pests and frost. This may be due also to aluminium ions released in the soil by acid rain and taken up by the roots of the trees. Figure 13.2 shows acid-rain damage to fir trees: they first lose their young leaves, later lose all their leaves and eventually die. The effect on humans of breathing sulphur dioxide gas is described in Section 18.5.

(b) Oil

In 1967 a giant oil tanker ran on to rocks off the coast of Cornwall and spilled 30 000 tonnes of oil into the sea. The oil polluted 100 miles of Cornish beaches and later, when the wind changed, polluted the coast of Brittany. Attempts to disperse the oil with detergent were unsuccessful. Ten days after the shipwreck, the ship was bombed from the air and the remaining oil burnt. This was the first

156   Figure 13.2   Acid-rain damage to trees

of many major oil escapes. Since then there have been other sinkings of giant oil tankers and oil slicks have polluted the sea around oil platforms.

Oil on the surface of water prevents diffusion of oxygen into the water with the result that many small animals die. It kills fish by clogging their gills and sea birds by clogging their feathers. On the shore it covers and kills animals and plants. Oil leaking from ships and oil released from ships cleaning their tanks at sea is now a major pollutant on many beaches in the world. The small black globules of oil on beaches are not harmful to swimmers but are a nuisance when they get on clothes and on the skin. Detergents used to disperse oil do more harm than good (see below).

(c) Radioactive waste
The atoms of most elements are stable but a few, such as those of uranium, are unstable and give off electromagnetic rays called **radiation**; anything giving off radiation is **radioactive**.

Figure 13.3 shows that the average person in Britain gets most radiation from natural sources: from the earth, from his or her own body and from outer space. For example, radon is a radioactive gas given off by rocks, such as granite, which contain uranium. Human activities cause about 13 per cent of the radiation received by humans. Most of it comes from the medical use of X-rays. About 0.1 per cent comes from gas and water discharges from industry, particularly from the nuclear-reprocessing plants at Sellafield and Dounreay. By reprocessing, spent (used) nuclear fuel is separated into 3 per cent highly active dangerous liquid waste, 1 per cent plutonium and 96 per cent reusable uranium.

Data given in Figure 13.3 are for the *average* person: data for individuals vary. People working in a nuclear-reprocessing plant, people working with X-rays or being given regular X-rays and people living over granite in Cornwall will all get above-average radiation.

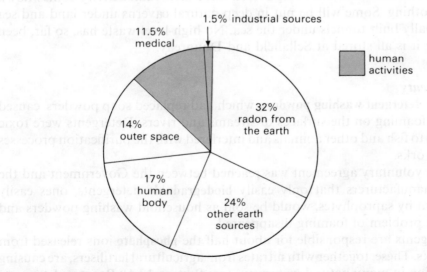

Figure 13.3 Radiation sources

157

None of our senses can detect radioactivity. A **Geiger counter** is an instrument that can. Radiation that can be absorbed by human tissues is measured in millisieverts (mSv). Most people get about 1 mSv a year from natural sources; someone exposed to radon could get 100 mSv a year; the maximum legal dose for radiation workers is 50 mSv a year. The firemen who put out the fire at the Chernobyl nuclear-power station, and who died within a few weeks of the accident, received doses of several thousand mSv.

No radiation dose is safe. It is estimated that 4 per cent of all cancers are caused by natural radiation. Most of what we know about the risks of radiation comes from data collected on the victims of the atom bombs dropped on Hiroshima and Nagasaki in 1945. Scientists at first underestimated the dangers of radiation. In 1987 the National Radiation Protection Board suggested that someone getting the maximum legal radiation limit (50 mSv a year) has a one in 700 chance of getting a fatal cancer.

Radioactive waste comes from the industrial uses of radioactive elements. It may be

○ **low-level**, for example, paper towels, rubber gloves, crushed glassware, protective clothing and chemical sludge from nuclear-power stations
○ **intermediate-level**, for example, solid and liquid materials from nuclear power stations and reprocessing plants
○ **high-level**, for example, the highly active liquid waste from nuclear reprocessing plants which is so dangerous that it must be artificially cooled and stored (at Sellafield and Dounreay) for at least 50 years

All these wastes become less dangerous in time because, as they give off radiation, their radioactivity gets less.

The UK Government has given UK Nirex Ltd, reorganised in 1985, the job of disposing safely of low-level and intermediate-level wastes. Some have been cast in concrete, enclosed in steel drums and dumped in the deep Atlantic ocean in an internationally agreed zone. Some have been isolated from the environment by concrete and steel and are being stored until their radioactivity comes down to almost nothing. Some will be put in deep natural caverns under land and sea and in specially built tunnels under the sea. No high-level waste has, so far, been disposed of: it is all stored at Sellafield and Dounreay.

(d) Detergents

In the 1950s **detergent washing powders**, which had replaced soap powders, caused widespread foaming on the surfaces of streams and rivers. Detergents were toxic (poisonous) to fish and other animals and interfered with the purification processes at sewage works.

In 1964 a voluntary agreement was reached between the Government and the detergent manufacturers that only easily **biodegradable** detergents, ones easily broken down by saprophytes, would be sold as household washing powders and liquids. The problem of foaming disappeared.

But detergents are responsible for about half the **phosphate** ions released from sewage works. These, together with nitrates from agricultural fertilisers, are causing eutrophication in many natural waterways (see Section 13.4). Because detergents are toxic (poisonous) to fish and other animals, their use to disperse oil at sea and on beaches does more damage to wildlife than the oil itself.

(e) Pesticides

Pesticides are intended to be harmful to pests (see Section 12.3). Very few of them are so **selective** that they harm pests and nothing else. Provided that they break down quickly, the pollution they cause (the harm they do to organisms other than pests and to the environment) is local and short-lived.

Herbicides are pesticides that kill plants. Hormone weedkillers are herbicides that kill broad-leaved weeds by upsetting their growth (resulting in an increase in the number of saprophytes feeding on dead weeds). Worms and mites in the soil are not affected by hormone weedkillers, which soon break down and seem to be harmless unless sprayed on to plants with broad leaves. But Agent Orange, which was spread from aircraft during the Vietnam War and which contained herbicides in high concentrations, has caused long-term ecological damage by killing trees and destroying mangrove swamps and has resulted in births of deformed babies.

Severe damage is done by pesticides that are **persistent** (last a long time) and are carried by animals or by water through the ecosystem. Some of the persistent insecticides, for example DDT, have done and are still doing widespread damage to the environment. DDT is everywhere in the world, in the air and in rain water (see Section 12.3).

13.4 Fertiliser pollution

In eastern England and in the Midlands a special form of pollution is causing concern to the water authorities responsible for supplying safe water to homes and industrial users. These regions of England grow vast quantities of cereal crops. Yields of these crops have greatly increased, partly because of the new varieties that are grown and partly because of the increased use of nitrogen fertilisers on which these new varieties depend.

Ammonium and nitrate ions from the fertilisers applied to these crops dissolve in rain water and get washed into the soil and into streams, rivers, lakes and underground water supplies. Streams and rivers were already polluted by ammonium, nitrate and phosphate ions from sewage works, and this source of pollution has been increased by the use of detergents. Figure 13.4 shows that in the 1970s the nitrate levels in the River Lee (in Essex) had risen above the World Health Organisation safe-level recommendation. Levels higher than this have since been recorded in rivers elsewhere in the UK.

A link between high-nitrate levels and an above-average level of 'blue-babies' is suspected but not proved. A health risk in adults is that nitrates may be chemically converted in the body to nitrites. It is suspected, but again has not been proved, that nitrite ions cause stomach cancer. In fact figures show that there is less stomach cancer in some areas of high-nitrate levels than in some areas of low-nitrate levels.

Natural water with high levels of inorganic ions is described as **eutrophic** and

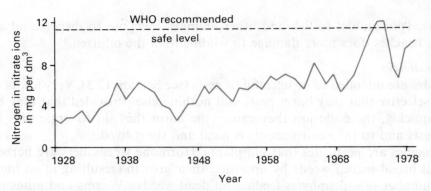

Figure 13.4   Nitrate levels in the River Lee (source: Royal Society Study Group, 1983)

the process of adding inorganic ions to water is **eutrophication**. Eutrophication is a natural process where rivers and lakes receive water that has drained through soils where salts are released in the natural inorganic-ion cycles. Natural eutrophication produces a balanced ecosystem in rivers and lakes. But inorganic ions from fertilisers and sewage have caused extreme eutrophication of many waterways. This has resulted in excessive plant growth, especially of microscopic algae (plant-like protoctists), at the surfaces of lakes, slow-flowing rivers and even of the sea where these rivers enter.

Concentrations of microscopic algae, resulting from eutrophication, are called **algal blooms**. Although they provide more food for animals, the algae grow and multiply so quickly that they cut off light from algae below them, and algae cannot make their food without light. The algae below them die and become food for saprophytes, which grow and multiply quickly. The saprophytes' respiration uses up the dissolved oxygen in the water. Animals, including fish, die from lack of oxygen. Eventually the water can become lifeless and smelly.

In summer even the large North Sea is affected: warm temperatures help algal growth and calm seas reduce wave action that reoxygenates the water. In May 1988, large numbers of fish in the North Sea and valuable crab and lobster stocks off Norway died from lack of oxygen.

Questions

Q 13.1. Table 13.1 shows deaths in various industries in one year in the UK. Why may these figures give a false impression of safety in the nuclear industry?

Q 13.2. In 1987 in Tolo Harbour, Hong Kong, a 'red tide' killed 120 tonnes of fish which would have been worth £24 000 to the local fishermen. A red tide is an algal bloom caused by a sudden vast increase in small plant-like red algae. The first red tide occurred in the summer of 1971, the next in 1975; there were six more in the late 1970s and there have been many more in the 1980s. Use the information in this passage and your own knowledge to suggest answers to the following questions.

Table 13.1   Deaths in one year in the UK

Industry	Deaths per 10 000 employees
Offshore oil and gas	15
Deep-mined coal	2.5
Oil-refining	1.5
Gas	0.7
Nuclear	0.14

(a) What killed the fish in Tolo Harbour?

(b) Why are red tides becoming more common?

Q 13.3. The French have developed a system that reduces the nitrate content of domestic water. Under anaerobic conditions they pass the water through a reactor containing straw, denitrifying bacteria and some simple inorganic compounds.

(a) What are (i) **anaerobic conditions** and (ii) **denitrifying bacteria**?

(b) Suggest reasons why (i) straw and (ii) simple inorganic compounds are included.

(c) How do you think this system helps to reduce the nitrate content of the water?

PART 2
Human and Social Biology

HUMAN CELLS

14.1 Cells

Unit 2 describes very small organisms, called microorganisms, such as bacteria and yeast. The basic unit of living material is a cell and many microorganisms have just one cell. A bacterial cell is not much more than cytoplasm, a long strand of DNA, a cell-surface membrane and an outer wall. A yeast cell is not much more than cytoplasm, a nucleus, a large vacuole, a cell-surface membrane and an outer wall.

Unit 8 describes plant cells from a leaf. A mesophyll cell has green chloroplasts as well as cytoplasm, a nucleus, a large vacuole, a cell-surface membrane and an outer wall. Plant cells are much bigger than those of microorganisms, and a flowering plant is made up of millions of cells.

Animal cells are bigger than those of microorganisms, smaller than those of plants. There are organisms of just one cell, but a new-born baby has about six thousand billion cells.

Because animal cells have no walls, it is not so easy to see where one animal cell ends and the next begins. The surface of an animal cell is covered by a very thin membrane, called the **cell-surface membrane**, which stops its contents spilling out. There are animal cells of different shapes and sizes: you will see a few of them. Inside an animal cell there is usually a **nucleus**, containing the important **chromosomes** with their **DNA**, surrounded by a **nuclear envelope**. Figure 14.1 shows the **cytoplasm**, nucleus, nuclear envelope and cell-surface membrane of an animal cell.

Unlike a plant cell, an animal cell has no cell wall and no large vacuole. (Animal cells may contain small vacuoles.) A plant cell's wall gives it a rigid shape which, together with the vacuole, supports the plant; animals that need support have

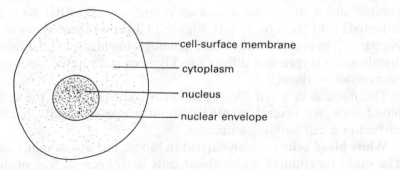

Figure 14.1   Animal cell

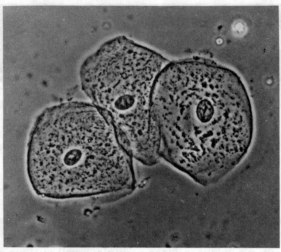

Figure 14.2   Human cheek cells (× 750)

skeletons. Many plant cells have chloroplasts, which are used in photosynthesis; no animal photosynthesises and no animal cell has chloroplasts.

(a) Human cheek cells

Like all other skin cells, cheek cells on the inside of the mouth flake off and are all the time being replaced. The practical work at the end of this Unit tells you how to use a microscope to see and draw some stained cheek cells. Figure 14.2 is a photograph of stained human cheek cells; you can see the cytoplasm and nuclei clearly; each cell is flat but has a different shape.

(b) Human blood cells

Cells are of different shapes and sizes which fit them to the functions they perform. Few are as straightforward as cheek cells. Most blood cells contain a sticky red protein pigment called **haemoglobin**, which gives blood its colour. The haemoglobin in **red blood cells** can combine with oxygen to form **oxyhaemoglobin**. It is as oxyhaemoglobin in red blood cells that oxygen is carried to all parts of the body through the blood vessels.

Red blood cells must be able to pick up oxygen easily wherever there is plenty and give it up easily wherever it is needed. A large surface area in relation to their volume and a shape like a licorice Pontefract cake (flat with each face deeply indented) help them to do this. Figure 14.3 shows three views of a red blood cell. No part of the cytoplasm, or of the haemoglobin inside it, is far from the cell-surface membrane. Oxygen can diffuse quickly because it never has to diffuse far to get into or out of the cell.

The nucleus in a red blood cell breaks up soon after the cell is formed. Red blood cells live several months without a nucleus: more haemoglobin can be carried in a cell without a nucleus.

White blood cells are also carried in blood, but not as many as red blood cells. The main function of white blood cells is to keep us free of disease. One type helps to make cells that can give out **antibodies** which kill bacteria and destroy

166

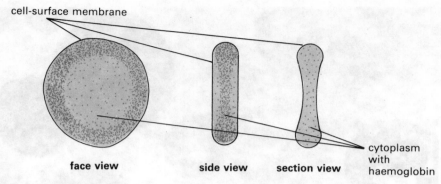

cell-surface membrane

face view   side view   section view

cytoplasm with haemoglobin

Figure 14.3   Red blood cells

viruses. Another type itself kills bacteria by surrounding them with cytoplasm and digesting them. Each white blood cell has a nucleus that helps to control its anti-disease activities. Figures 14.4 and 14.5 show three different types of white blood cell; the ones with lobed nuclei (the nuclei that are not round) capture and kill bacteria.

Figure 14.5 is a photograph of several red and white blood cells. The red ones have dark rims and light centres because there is more cytoplasm and haemoglobin around the rims and less at the centres where they are indented.

(c) Human gametes

The cells that meet and fuse with one another to form a new offspring during sexual reproduction are called **gametes**. The female gamete or **egg cell** is a large simple-looking cell, much like the one in Figure 14.1. Its nucleus is important because it contains the DNA that provides the half of the new offspring's inherited material which comes from the mother. Its cytoplasm contains food for the first stages in the growth of the offspring. It is surrounded by a thick outer coat which only a healthy strong sperm can get through. Figure 14.6 shows a human egg cell and a sperm just before they fuse. (The two small cells at the edge of the egg cell are formed in the process of removing half the DNA of the mother's cell.)

The male gamete or **sperm** is different from the egg cell. Much smaller, it has a head containing the nucleus with its half of the DNA for the new offspring. Its cytoplasm is stretched out in a long **tail** with which it swims. To reach an egg cell, sperms have to swim what is a long way for their size. Just behind the nucleus, in the middle part, the cytoplasm contains **mitochondria**, special packages

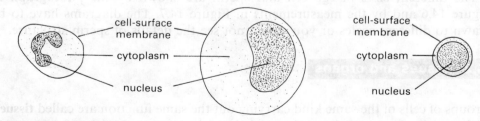

cell-surface membrane

cytoplasm

nucleus

cell-surface membrane

cytoplasm

nucleus

Figure 14.4   White blood cells

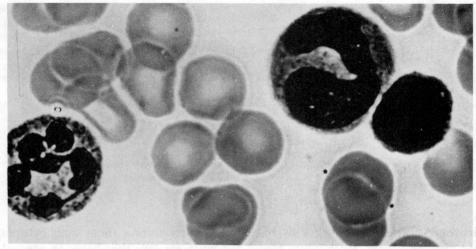

Figure 14.5   Blood cells (× 2000)

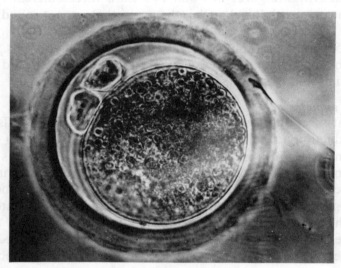

Figure 14.6   Human egg cell and sperm (× 360)

containing enzymes, which help release the energy for the long journey that sperms must make. Food for the journey comes from the liquid called **semen**, in which they are released, and from the cells that line the passages through which they swim.

The different sizes of egg cell and sperm are shown by the photograph in Figure 14.6 and by the measurements in Figure 14.7. The diagrams have to be drawn to different scales or you would not see the details of sperm structure.

14.2 Tissues and organs

Groups of cells of the same kind carrying out the same function are called **tissues**. 168 Different tissues grouped together to help with the same function form **organs**.

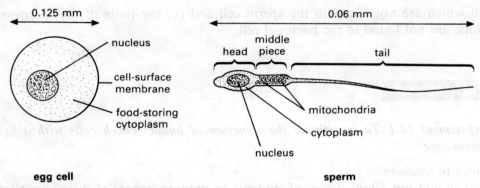

egg cell

sperm

Figure 14.7   Human egg cell and sperm (diagram)

For example, the cheek cells you will see in the practical work are part of the tissue called **epithelium**. Together with other tissues they form the organ called the **skin**. The heart is an organ covered on its inner and outer surfaces by epithelium but consisting mainly of **muscle** tissue. It contains another tissue, blood. The brain is an organ surrounded by epithelium but containing mostly **nerve** tissue. Our bodies contain cells grouped into tissues grouped into organs.

Plants too have cells, tissues and organs. For example, the leaf is an organ made up of tissues, such as mesophyll and epidermis, which consist of cells.

Questions

Q 14.1. Describe *three* human cells and explain how they are suited to the functions they carry out.

Q 14.2. Draw a yeast cell, a mesophyll cell and a cheek (epithelial) cell next to one another. Label their parts. Make a table like the one shown below. Put a tick when a cell has the part listed in the left-hand column and a cross when it does not. The first line has been done for you.

	Yeast cell	Mesophyll cell	Cheek cell
Cell wall	✓	✓	✗
Cell-surface membrane			
Cytoplasm			
Nucleus			
DNA			
Vacuole			
Chloroplast			

Q 14.3. Look at the bacterial cell in Figure 2.3 and the sperm cell in Figure 14.7. Make a list of (a) the parts they have in common, (b) the parts of the bacterial   169 cell which are not found in the sperm cell and (c) the parts of the sperm cell which are not found in the bacterial cell.

Experiment 14.1 *To investigate the structure of human cheek cells with a light microscope*

Notes to teachers
You should not allow classes of students to prepare smears of their own cheek (epithelial) cells. Commercially prepared slides are available from biological suppliers or you may substitute photographs of cheek cells.

Materials needed by each student
1 microscope
1 permanent stained preparation of human cheek cells

Instructions to students
1. Put the microscope on a firm surface. Depending on the type of microscope you have, either switch on the built-in light or adjust the mirror to reflect as much light as possible into the microscope. Turn the nosepiece of the microscope so that the lowest-power objective lens clicks firmly into place.
2. Clip the slide on the stage of the microscope so that the edge of its label is directly below the objective lens.
3. Turn the coarse-focus screw of the microscope until the bottom of the objective lens is as close to the slide as it will go. So long as you are using the low-power objective lens there is no danger that it will hit the slide. Looking down the microscope, turn the coarse-focus screw until the label of the slide comes into focus.
4. Move the slide about until you find some stained cells. They are usually in groups and will appear darker than the unstained surrounding areas. Ignore any darkly stained strands that are just bits of debris which happened to be on the slide as it was being prepared.
5. Move the slide until a cell is directly in the middle of your field of view. Turn the nosepiece of the microscope until a higher-power objective lens clicks firmly into place. You must take great care not to let this objective lens touch the slide. While looking at the objective lens from the side, turn the coarse-focus screw of the microscope until the bottom of the objective lens is as close to the slide as it will go. When you look down the microscope at this magnification, always rack upwards. Now you can re-examine the cells. Use the fine-focus screw if you need to re-focus.
6. Make a large accurate drawing of one of the cells you see. Appendix C tells you how to draw biological material. As a general rule you should draw and label only what you see: you should not draw and label what you think ought to be there. In this experiment you will see and draw a line where the cytoplasm ends. The line you see is simply the surface of the cytoplasm. The cell-surface

170

membrane is in the same place as the line but is too thin for you to see. You may, however, label the line you draw 'cell-surface membrane'.

Interpretation of results

1. The cheek cells have been stained so that their structure appears clearer. Stains react with chemicals in cells, colouring them as a result. Suggest why different parts of the cell have stained differently.
2. Label the **nucleus** in your drawing. Suggest why you cannot see any chromosomes in it.
3. Label the **cytoplasm** in your drawing. Can you see any structures in the cytoplasm? If so, are they the same colour as or a different colour from the nucleus?
4. Why have you not been able to see the cell-surface membrane at the edge of the cytoplasm?
5. Why do you think you were told to focus the microscope on the label of the slide before looking for cells?

FOOD

15.1 Introduction

Most food is of three kinds: **carbohydrates** and **lipids** (fats and oils) both provide energy; **proteins** are used for growth and for repair of tissue. We also need **inorganic ions** and **vitamins** for good health. Every day too we should drink **water** and take some of our carbohydrate in the form of **roughage**, which prevents constipation.

15.2 Carbohydrates

Carbohydrates are the food stores which most plants draw on for their own energy. (Some seeds store lipids.) They are made up chemically of only three elements, **carbon**, **hydrogen** and **oxygen**, in the appproximate ratio of 1:2:1. You have met some carbohydrates in earlier units: **glucose** ($C_6H_{12}O_6$), the sugar used in all cells to release energy in respiration; **cellulose**, the substance that forms plant cell walls; **starch**, the substance stored in plant cells; **glycogen**, the substance stored in yeast cells.

Animals like ourselves also store carbohydrate as glycogen in muscles and in the liver. But animals store food mainly as lipid and store only small amounts as carbohydrate (glycogen). It is in plants that carbohydrate (starch) is the main storage product.

Molecules of simple sugars, such as glucose, become linked to one another if a molecule of water (H_2O) is formed by taking a hydrogen ion (H^+) from one of them and a hydroxyl ion (OH^-) from the other. This is shown diagrammatically in Figure 15.1. Even a simple sugar has a complicated chemical formula of which OH^- is a small part.

When two simple-sugar molecules are joined, a complex-sugar molecule is formed. Cellulose, starch and glycogen consist of thousands of simple-sugar molecules joined to form unbranched (cellulose) and branched (starch and glycogen) chains.

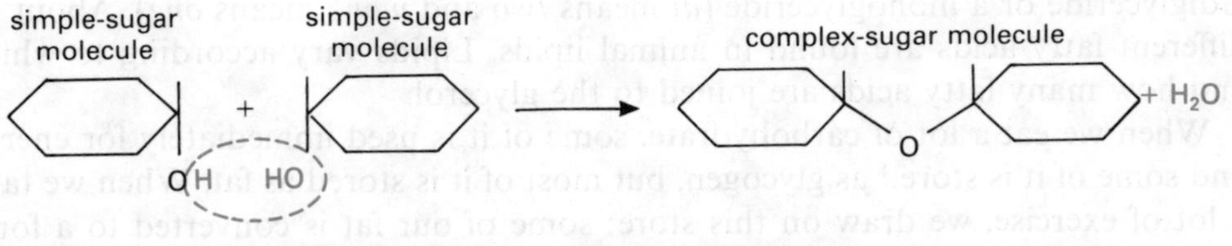

Figure 15.1   Linkage of two simple-sugar molecules

173

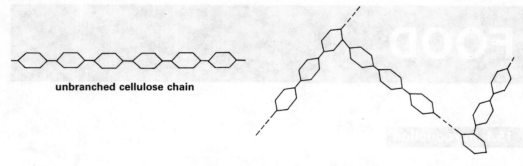

unbranched cellulose chain

branched starch or glycogen chain

Figure 15.2 Linked molecules of simple sugars

(a) Sources of carbohydrate in the diet

Plant products are rich in carbohydrates. They usually contain starch. Crops of cereal grains supply the starch we get from different flours, of which wheat flour is the most important. Sugar comes in vast quantities from sugar-cane and sugar-beet and in small quantities from fruit. All plants contain cellulose, which we cannot digest and is therefore no use to us as food. Because we cannot digest it, however, cellulose gives the gut bulk to work against and is our main source of the roughage that prevents constipation. We get almost none of our carbohydrate from animal products because they contain so little.

15.3 Lipids

Lipids, commonly known as **fats** when they are solid and **oils** when they are liquid, are the food stores which animals rely on for energy. We store lipids as fat under the skin and around organs such as the kidneys. Lipids are also present in every living cell as part of the cell-surface membrane. Fat under the skin helps to insulate us by slowing down heat loss from the body. Long-distance swimmers, who are more likely to fail from cold than from exhaustion, start well insulated with their own fat stores. For the same mass, lipids are a better energy source than carbohydrates. Like carbohydrates, lipids are made up of only the three elements, **carbon**, **hydrogen** and **oxygen**, but they contain a smaller proportion of oxygen. Like complex carbohydrates, lipids are made up of smaller units joined to one another by the formation of molecules of water. Figure 15.3 shows how a lipid results from three **fatty acids** and a molecule of **glycerol** with the formation of three molecules of water. The lipid formed in this way is a **triglyceride** (*tri* means *three*). Glycerol can also join with only two or even only one fatty acid to form a diglyceride or a monoglyceride (*di* means *two* and *mono* means *one*). About 30 different fatty acids are found in animal lipids. Lipids vary according to which and how many fatty acids are joined to the glycerol.

When we eat a lot of carbohydrate, some of it is used immediately for energy and some of it is stored as glycogen, but most of it is stored as fat. When we take a lot of exercise, we draw on this store: some of our fat is converted to a form in which it can be used up in respiration to give us energy.

174

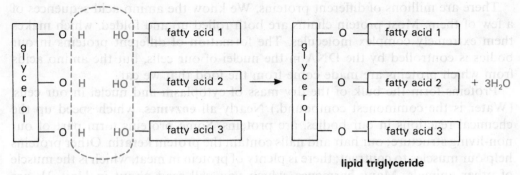

Figure 15.3   Linkage of glycerol and three fatty acids

(a) Sources of lipid in the diet

Although we can convert carbohydrate into fat, we need certain lipids in our diet: these contain **essential fatty acids**, essential because we cannot make them from carbohydrates. Lipid is such a rich source of energy that in the UK most of us eat far more than we need: we eat the fat that surrounds the muscles in meat and fat in the forms of butter and margarine; we also eat oils in which we cook our food.

15.4 Protein

Because proteins are used for both growth and repair of tissue, everyone needs them, but children need much more of them (in proportion to their size). Because most proteins come from animal products, it is probably better not to become a vegetarian or a vegan until you have finished growing. A vegetarian eats no meat but does eat eggs and dairy products (milk, butter and cheese). A vegan eats no animal products at all. If you are a vegetarian or a vegan, and especially if you are a vegan, you should eat lots of pulses (peas, lentils and beans), which are also rich in protein.

Like complex carbohydrates and lipids, proteins are built up from simpler units. These units are **amino acids**. All amino acids contain the same three elements as carbohydrates and lipids (carbon, hydrogen and oxygen) plus one more, **nitrogen**; two of the amino acids also contain **sulphur**. Although only twenty amino acids are found in living organisms, they can be linked in chains, by the formation of a molecule of water from each pair, in their hundreds and thousands in an almost infinite number of different combinations. This linkage is shown in Figure 15.4.

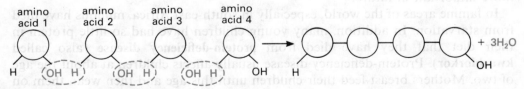

Figure 15.4   Linkage of amino acids

There are millions of different proteins. We know the amino-acid sequences of a few of them. Most protein chains are both rolled up and folded, which makes them extremely complex molecules. The formation of different proteins in our bodies is controlled by the DNA in the nuclei of our cells, but the amino acids from which proteins are made come from the food that we eat.

Proteins form the bulk of the dry mass of cytoplasm and nuclei in our cells. (Water is the commonest compound.) Nearly all **enzymes**, which speed up the chemical reactions in our bodies, are proteins. Some proteins form part of our non-living structure; our hair and nails contain the protein **keratin**. Other proteins help our muscles to contract: there is plenty of protein in meat, which is the muscle of other animals. Many **hormones**, which you will read about in Unit 21, are proteins. **Haemoglobin**, the red pigment in red blood cells, is a protein. **Antibodies**, the substances that protect us against disease, are proteins. **Fibrinogen**, a substance that helps our blood to clot and stops us bleeding to death, is yet another protein.

Because proteins contain nitrogen and sometimes sulphur and even other elements (for example, haemoglobin contains iron), they cannot be made up from carbohydrates and lipids. Protein in our diet gives us the amino acids and other elements from which we can build up our own proteins. Our own proteins also contain elements from other sources such as inorganic ions in cereal products and vegetables.

Because protein cannot be stored in the body, we need a small but continuous supply of it to stay healthy: it is best to eat some protein every day. About 15 per cent of the body mass consists of protein. Protein is needed for growth and for repair of tissue. Pregnant women and women feeding their babies on their own milk need more protein.

Protein itself cannot be stored and only a proportion of it can be converted to lipid and stored. For this reason protein is the least fattening of the basic foods: people who are slimming often eat more protein and less carbohydrate and lipid.

Although protein cannot be stored, it does contain the elements from which carbohydrates and lipids are made. When people are starving, the proteins forming their bodies are broken down so that the carbon, hydrogen and oxygen in them can be used for the release of energy. It is because starving people feed on their own bodies in this way that they are left looking like skin and bones.

(a) Sources of protein in the diet

Animal products are an excellent source of protein: meat, fish, eggs, milk and products of milk such as cheese and yoghurt. Plant products, especially seeds, which provide the protein for growing plants, also supply protein in our diet: peas, lentils and beans are rich in protein. **Single-cell protein (SCP)** and mycoprotein (see Section 7.6) are other sources of protein which will surely become more important in future.

In famine areas of the world, especially in south-east Africa, millions have died from starvation. In addition, many young children have had so little protein in their diet that they have died from **protein-deficiency disease** (also called **kwashiorkor**). Protein-deficiency disease usually affects children at about the age of two. Mothers breast-feed their children until this age and then wean them on whatever food they can get, which may be only a dilute porridge. The protein

176

value of such porridge is far too low for growing children. A child with protein-deficiency disease has little muscle, its liver swells and its abdomen is swollen further because the tissues retain water. Because of their weakness and poor state of health, children in famine areas die not only of starvation and protein-deficiency disease but of other diseases such as measles and dysentery.

15.5 Inorganic ions

We need a number of different **inorganic ions** in our diet. Two very important ones are **iron** and **calcium**.

(a) Iron
More than half the iron in our bodies is in the protein haemoglobin, the red pigment in blood; some iron is present in another red pigment in our muscles. We can store iron in the liver. Not much iron is lost except by bleeding after an accident and by women during **menstruation**, the monthly loss of blood when the lining of the uterus is shed.

Most adults keep healthy on a mere 10 mg of iron a day. Growing children need more because they are making more blood and muscle; a pregnant woman needs more for the baby growing inside her; anyone who has lost blood needs more. Because a new-born baby has a store of iron in its liver, it does not matter that milk contains little iron.

(i) Sources of iron in the diet
Meat products contain iron: the best source is liver because that is where iron is stored. Plant products also contain iron: cocoa powder is almost as good a source as liver. Flour sold in the UK has to contain a certain amount of iron by law; if not enough is left after the grain is milled into flour, iron compounds are added to it.

(b) Calcium
Calcium is needed by the body in far greater amounts than iron. Calcium is the most abundant mineral in our bodies; almost all of it is in our bones and teeth and helps to harden them. Small amounts of calcium are also needed for the contraction of muscles, the passing of nerve impulses, the activity of some enzymes and the clotting of blood. As you would expect, growing children, pregnant women and women feeding their babies on their own milk need more calcium than other people.

It may surprise you that older people need more calcium too. In older people, especially women, bones tend to become brittle for lack of enough calcium and to break easily. The explanation is that throughout life, as calcium is needed elsewhere in the body, it is taken from bones. In other words, bones act as a store of calcium; and in older people calcium is removed faster than it is replaced. We do not understand why this should happen. All we known is that the greater brittleness of women's bones has something to do with lack of hormones after the menopause. (In the UK more women die of **brittle-bone disease** than of cancers of the breast, cervix and uterus put together.)

177

(i) Sources of calcium in the diet

Milk and cheese are excellent sources of calcium: you should expect milk, and therefore cheese, to be rich in calcium because a young mammal may grow entirely on milk for the first few months of its life. Because in the UK white flour must contain a certain amount of calcium by law, calcium compounds (such as chalk) may, like iron compounds, have to be added to flour after milling. The 1963 Bread and Flour Regulations control what is allowed and what is essential in bread and flour.

15.6 Vitamins

Vitamins are substances needed in only small amounts to keep us healthy. They were not discovered until the beginning of this century, but long before then sailors knew that fresh fruit and vegetables prevented **scurvy**. Now we know that scurvy is prevented by a small amount of vitamin C, which is present in most fruit and vegetables. Vitamin C gives us a healthy skin and allows wounds to heal quickly; it helps in the absorption of iron in the gut; it may have other good effects, such as helping to prevent us catching colds.

Vitamin D is needed for the development of strong bones. It acts by helping the absorption of calcium (which makes bones hard) from the gut. Too much vitamin D is dangerous because too much calcium damages the kidneys. We make vitamin D in our own bodies by the action of sunlight on a substance in our skin. Even in places, such as northern Europe, where there is not a lot of sunlight, unpigmented skin can usually absorb enough sunlight to make enough vitamin D; but those (such as the elderly) who stay indoors, those with dark skins and those who wear a lot of clothes when they go out all run a risk of vitamin D deficiency (too little vitamin D).

Vitamin D deficiency is very noticeable in children: their legs become bowed or bandy because the bones in them are too soft and weak (from lack of calcium) to support their weight. The disease they are suffering from is called **rickets**. Figure 15.5 is an X-ray photograph of the legs of a child with rickets: you can see that the lower-leg bones are bent above the ankle.

Apart from the action of sunlight on our skin, the only natural sources of vitamin D are animal products, especially fish oils. By law, vitamin D is added to margarine, which helps preserve the health of vegetarians. It is also added to chapati flour.

15.7 Water

About 60 per cent of our body mass is water. It forms the major part of cytoplasm and nuclei and the major part of blood. Water is the substance in which other chemicals are dissolved, in which chemical reactions take place and in which materials are transported. During digestion water reacts with carbohydrates, lipids

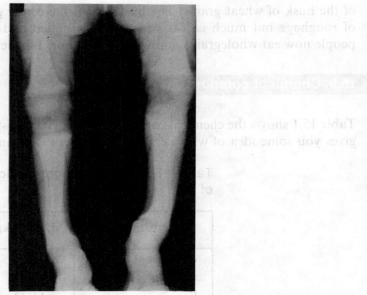

Figure 15.5   Legs of a child with rickets

and proteins to reverse the reactions seen in Figures 15.1, 15.3 and 15.4 to form simple sugars, glycerol, fatty acids and amino acids.

Every day we lose a certain amount of water from our bodies. It goes in urine, in sweat, in breath, in faeces and in any other fluids, such as tears and blood, which leave the body. It also simply evaporates from the skin, even on a cold day when we do not sweat much. To make up for the constant loss of water we need water in our diet. Most of our food contains some water. If we do not have enough water in our bodies, a thirst centre in our brain makes us feel thirsty and we have an urge to drink.

15.8 Roughage

Dietary fibre, or **roughage** as it is commonly called, is indigestible material that we eat. It comes from plants and consists mostly of the carbohydrate **cellulose**. Even animals that live entirely on plant food, such as cows and horses, cannot themselves digest cellulose: they rely on bacteria living in their guts to digest it for them. (Bacteria in our guts probably digest a little cellulose for us.)

It has long been known that roughage helps move food quickly through the gut and that it prevents constipation: it gives the muscles in the wall of the gut something to work against, pushing the food through by a process called **peristalsis** (see Section 16.1). There is now evidence that dietary fibre may help to reduce the risk of heart disease, of bowel cancer and of gallstones.

(a) Sources of roughage in the diet

Since roughage comes from plant cell walls, all vegetables and fruits will contain some though not necessarily very much. A rich source is **bran**, which is made up   179 of the husk of wheat grains. The husks of all the cereal grains are good sources of roughage but much is lost when flour is milled and rice is polished. Many people now eat wholegrains and wholemeal flours for the roughage they contain.

15.9 Chemical composition of the human body

Table 15.1 shows the chemical composition of a 70 kg 25-year-old man. The table gives you some idea of why we need a variety of substances in our diet.

Table 15.1 Chemical composition of a 70 kg man

Substance	Mass in kg
Water	43.40
Protein	11.60
Fat	11.43
Glycogen	1.00
Inorganic ions:	
Calcium	1.27
Iron	0.01
Others	1.29

15.10 World food problems

We know that some parts of the world have huge surpluses of food while in other parts people are starving. Though it may sound an easy matter to move food from one part to another, anyone who has followed the attempts by Live Aid and other organisations to feed the starving in Africa knows there are no simple solutions. Transporting food in the quantities needed to places with no roads or airstrips is almost impossible.

The famines in Ethiopia illustrate more of the problems. The people in the north of Ethiopia, which is where the famine areas are, used to move south with their cattle between the months of September and April, which is when the rains move south, and north again between May and August, which is when the rains move north. The grass on which they grazed their cattle then had a few months in every year in which to recover.

In the 1930s new water holes were opened in the north which discouraged the people who lived there from spending September to April in the south. By 1985 the peanut-growing areas in the south had increased sixfold (possibly as the result of earlier Aid programmes), which meant it was no longer possible for the people in the north to graze their cattle there from September to April. The switch from a nomadic life to a settled agriculture was complete. The result was that the land in the north became overgrazed and the vegetation on it began to disappear.

You known from reading about the water cycle in Section 10.2 that water evaporates from vegetation and returns to the land as rain. In northern Ethiopia the remaining vegetation received less rain and the land eventually turned to dust. **Desertification** occurs because of a vicious circle: less vegetation means less rain, which means less vegetation, which means less rain, and so on. Both vegetation and rainfall in the north of Ethiopia have steadily declined over the last twenty years. The people in the north cannot even afford to buy the peanuts grown in the south.

What are the long-term solutions? Cattle herds could be restricted in number to make staggered grazing possible again, so that grass is always given time to recover. New water holes should not be dug where they will discourage staggered grazing. People in the desert areas of the north could be encouraged to go to new areas while their land recovers. The best way to speed recovery is to introduce carefully selected plants which bind the soil together, which survive with little moisture and which do not taste good to grazing animals. Plants that bind the soil together are known as **green glue**. The green glue would improve the rainfall, after which grass could grow again. Whatever is done, restoring the desert areas to grazing land is likely to take 25 years or more.

Questions

Q 15.1. Make a table like the one shown below but larger. Complete it to summarise what you have learnt in this Unit about food. The first line has been filled in for you.

	Carbohydrate	Lipid	Protein
Basic units Elements present Examples Functions in the body Where present in the body Sources in the diet	simple sugars	fatty acids	amino acids

Q 15.2. Make a table like the one shown below but larger. Complete it to summarise what you have learnt in this Unit about iron and calcium ions and about vitamins C and D.

	Iron	Calcium	Vitamin C	Vitamin D
Functions in the body Sources in the diet				

181

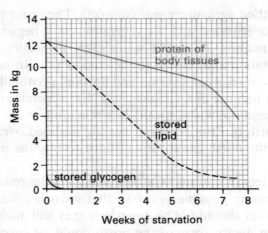

Figure 15.6   Food used up during starvation

Q 15.3. Figure 15.6 shows the amount of protein, lipid and glycogen which a 70 kg man used up when he starved in an experiment that lasted several weeks.

(a) How much of each of the three compounds had been used up by the end of the first week?

(b) How long did it take for (i) all the glycogen, (ii) half the lipid and (iii) 2 kg protein to be used up?

(c) Why do you think starvation beyond $5\frac{1}{2}$ weeks is dangerous? Use evidence from the graph to explain your answer.

Practical work

Experiment 15.1 To test for starch and reducing sugar

Note to teachers
Because the concept of a reducing sugar is difficult, only *simple sugars* and *complex sugars* are referred to in the text of this book. The LEAG syllabus requires the identification of starch and reducing sugar but no understanding of what a reducing sugar is. You may like to explain to some students the distinction between reducing and non-reducing sugars and to demonstrate how a non-reducing sugar such as sucrose can be hydrolysed by boiling it with dilute hydrochloric acid to produce reducing sugars. (Cool and neutralise the boiled mixture using sodium carbonate before adding Benedict's solution and reboiling.)

Uncontaminated glucose solution will not give a positive result with iodine solution but, unless the starch suspension is fresh, starch may give a weak positive result with Benedict's solution.

Note to students
The simple sugars, such as glucose, are all reducing sugars. Some complex sugars are also reducing sugars; others, such as cane sugar, are non-reducing sugars. A positive Benedict's-solution test does not tell you if a simple or complex sugar is present: it tells you only that a reducing sugar is present.

Materials needed by each student
1 250 cm³ beaker
1 2 cm³ pipette fitted with a pipette filler
1 dropper pipette
1 test-tube rack with 2 test-tubes
1 dimple tile
1 Bunsen burner, tripod and gauze
1 pair of tongs (to hold hot test-tubes)
1 spirit marker
10 cm³ Benedict's solution
5 cm³ iodine in potassium-iodide solution
5 cm³ 1% glucose solution
5 cm³ freshly prepared 1% starch suspension

Materials needed by the class
distilled water
tap water

Instructions to students
1. Copy the table.

Nature of solution	Colour of solution
Iodine solution Iodine solution with added starch Iodine solution with added distilled water	

2. Put single drops of iodine solution in two wells of the dimple tile. Record the colour of the iodine solution on the first line of the table.
3. Add one drop of starch suspension to one of the drops of iodine solution and one drop of distilled water to the other drop of iodine solution. Record the colour of the mixtures on the second and third lines of the table.
4. Copy the table.

Tube	Contents of tube	Colour before boiling	Colour after boiling
A B			

5. Half fill a beaker with tap water and make a water bath by boiling the water using the Bunsen burner, tripod and gauze.
6. While you are waiting for the water to boil, label the test-tubes A and B. 183

Pipette 2 cm^3 of Benedict's solution into each of these test-tubes. Pipette 2 cm^3 of glucose solution into test-tube A and 2 cm^3 of distilled water into test-tube B. Record the composition and colour of these mixtures in the appropriate spaces in your table.

7. Put the two test-tubes in the beaker of boiling water. After about a minute, record the colours of their contents in the final spaces of your table.

Interpretation of results
1. Copy the table and use it to record your results.

Chemical	Description of test	Description of positive result
Glucose (reducing sugar) Starch		

Make sure that when you describe positive results you give the colours at the beginning and end of the test. 'Colour changes from red to green' is a better description than 'goes green'.
2. Why was distilled water used in each of the two tests?
3. Why were you told to boil the contents of the test-tubes by putting them in a beaker of boiling water rather than by holding them in a Bunsen flame?

Experiment 15.2 To test foods for starch and reducing sugar

Note to teachers
Students must do Experiment 15.1 before they do this one.

Materials needed by each student
1 250 cm^3 beaker
1 2 cm^3 pipette fitted with a pipette filler
1 dropper pipette
1 test-tube rack with 2 test-tubes
1 dimple tile
1 Bunsen burner, tripod and gauze
1 pair of tongs (to hold hot test-tubes)
1 spirit marker
10 cm^3 Benedict's solution
5 cm^3 iodine in potassium-iodide solution

Materials needed by the class
pestles and mortars
distilled water
tap water
a selection of foods suitable for testing for starch and reducing sugar: for example, rice, potato, apple, onion

184

Instructions to students

1. Make a table in which you can record your results. Appendix A helps you design a table.
2. Choose a food. If it is not already a liquid, make a liquid extract of it by crushing it with a pestle and mortar and mixing it with a little distilled water.
3. Test this extract for the presence of starch and reducing sugar using the tests you learned in Experiment 15.1.
4. When you have done your experiment, write a report telling what you did, what the results were and what these show about the composition of the food (or foods) which you tested. Appendix B helps you write a report.

EATING AND DIET

16.1 The gut

When we eat, we take food in at the mouth, chew it and swallow it. This is how food enters the gut, a continuous tube which includes the **stomach**, the **small intestine** and the **large intestine** and which goes all the way from the mouth to the **anus**. Some food is absorbed into the body from the gut. What is not absorbed is got rid of from the anus as **faeces**.

The gut has a wall of two sets of muscles. One set, the **longitudinal muscles**, lies lengthwise along the gut; the other set, the **circular muscles**, lies around the gut. These muscles are not normally under our conscious control; most of the time we do not even know they are at work. When muscles in any one set contract (shorten), those level with them in the other set relax.

When the circular muscles behind food contract, the longitudinal muscles level with them relax. The effect is to push the food along the gut. At the same time the longitudinal muscles in front of the food contract and the circular muscles level with them relax: the effect is to pull the gut up to the food. This method of moving the food along the gut is known as **peristalsis** and is shown in Figure 16.1. Imagine moving a tennis ball inside a tight sock: you would push the ball forward by squeezing the sock behind it; then you would pull the sock up towards the ball; and so on.

Figure 16.2 shows the position of many of the organs of the gut in the human body; it does not show how one organ leads into another.

(a) Digestion

Food is a mixture of large insoluble molecules and small soluble molecules. During movement of food through the gut the large insoluble molecules in it are chemically

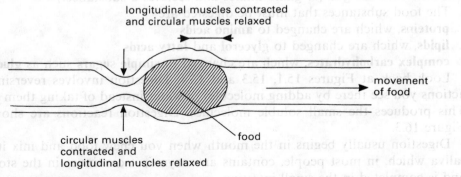

Figure 16.1   Food moved along the gut

187

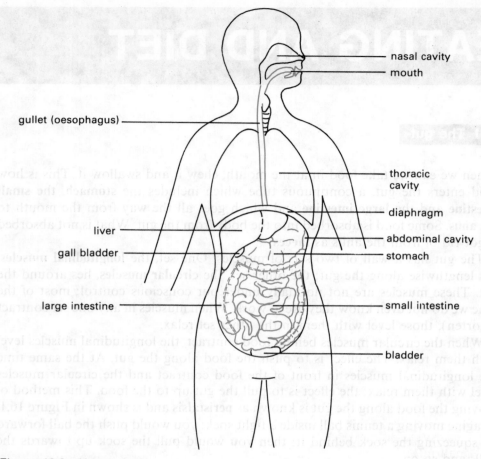

nasal cavity
mouth
gullet (oesophagus)
thoracic cavity
diaphragm
liver
abdominal cavity
gall bladder
stomach
large intestine
small intestine
bladder

Figure 16.2 Human gut

broken down, with the help of **enzymes**, to small soluble molecules. The small soluble molecules are absorbed through the walls of the gut and are transported to the cells where they are needed. This **digestion** inside the gut is the same process as the digestion of food outside the bodies of saprophytic fungi and bacteria which is described in Unit 7. The enzymes are made in the wall of the gut or in **glands** that lead into the gut. (A gland makes and secretes a substance.)

The food substances that must be digested are:

○ **proteins**, which are changed to **amino acids**
○ **lipids**, which are changed to **glycerol** and **fatty acids**
○ **complex carbohydrates**, which are changed to **simple sugars** such as **glucose**

Look back at Figures 15.1, 15.3 and 15.4. Digestion involves reversing the actions you see there by adding molecules of water instead of taking them away. This produces the small soluble molecules. Digestion reactions are shown in Figure 16.3.

Digestion usually begins in the mouth when you chew food and mix it with saliva which, in most people, contains an enzyme; it continues in the stomach and is completed in the small intestine.

188

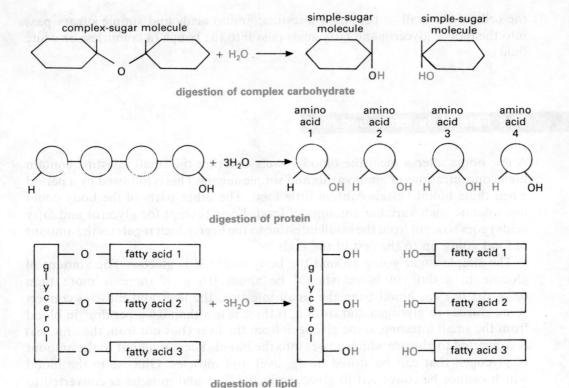

digestion of complex carbohydrate

digestion of protein

digestion of lipid

Figure 16.3   Digestion of proteins, lipids and complex carbohydrates

(b) Absorption

The small intestine is the longest organ of the gut. There is time as food passes through it for absorption of the digested food in the form of amino acids, glycerol, fatty acids and simple sugars. Absorption is helped by millions of finger-like projections, called **villi**, which cover the inner lining of the small intestine and greatly increase its surface area: the greater the surface area, the greater the rate of absorption. Villi are shown in Figure 16.4. The soluble food is absorbed through

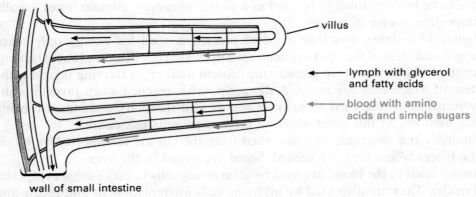

villus

lymph with glycerol and fatty acids

blood with amino acids and simple sugars

wall of small intestine

Figure 16.4   Villi

the cells of the wall of the small intestine: amino acids and simple sugars pass into the blood; glycerol and fatty acids pass into the **lymph** (a colourless or white fluid).

16.2 Fate of the absorbed food

A few hours after a meal the blood vessels covering the small intestine contain high concentrations of amino acids and simple sugars. This is followed by a period when these blood vessels contain little food. The other parts of the body could not tolerate such variable amounts of food. Food (except for glycerol and fatty acids) goes straight from the small intestine to the **liver**, which **regulates** the amount of food going on to the rest of the body.

The simple sugar going around the body in blood is **glucose**. The amount of glucose in a dm^3 of blood should be about 0.6 g. If there is more than 0.6 g per dm^3 in blood from the small intestine, the liver and muscles convert some glucose to glycogen and store it. If there is less than 0.6 g per dm^3 in blood from the small intestine, some glycogen from the liver (but not from the muscles) is converted to glucose which passes into the blood. There is a limit to the amount of glycogen that can be stored in the liver and muscles. Glucose in the blood which cannot be converted to glycogen in the liver and muscles is converted to fat and stored under the skin or around organs such as the kidneys. This is why you become fat if you eat a lot of sugar.

If there is too much of any particular amino acid, the liver converts the excess to **urea**, which passes into the blood and is later got rid of in **urine**.

Excess glycerol and fatty acids pass in lymph through the vessels of the lymphatic system before they are passed into the blood in the neck region. While in the **lymphatic system**, the glycerol and fatty acids are reconverted into minute droplets of fat. Excess fat is deposited under the skin or around organs such as the kidneys.

All body cells need a supply of food as a source of energy: they usually get it in the form of glucose from the blood. Glucose in the blood comes from food that has recently been eaten or from glycogen stored in the liver. Lipid stores may also be broken down to be used as a source of energy. Protein is not usually an important source of energy.

Figure 15.6 shows you that a lot of protein is used for energy only when glycogen and most of the lipid has been used up, which is when people are starving. Since the body cannot store protein, the protein used up in starving people is the protein of which cells are made. A danger point is reached when protein is the only substance being used to provide energy and the body's organs are being slowly used up. At this point *anorexia nervosa* patients risk death.

Vitamins and inorganic ions absorbed from the gut are transported by blood to the places where they are needed. Some are stored in the liver.

Amino acids in the blood are used by all growing cells to make more cytoplasm and nuclei. They are also used by all living cells for replacement and repair and especially for making enzymes.

190

The roughage in food remains undigested. By the time food reaches the end of the small intestine most of the digestible matter has been taken out. It now consists mainly of roughage, of a large number of bacteria that have been living and growing on the food and of a lot of water which has helped to dissolve the digested food.

From the small intestine food passes into the **large intestine**, where most of the water is now absorbed through the wall into the blood. Certain vitamins made by some of the gut bacteria may also be absorbed here. In the first part of the large intestine, the **colon**, undigested food and bacteria are made into **faeces**, which pass into the last part of the large intestine, the **rectum**, and finally pass out through the **anus** during **egestion** (defaecation).

16.4 Diet

A healthy diet has
o enough energy value in the food
o the correct balance between carbohydrates, lipids and proteins
o enough roughage
o various vitamins
o various inorganic ions

(a) Energy value
In the practical work at the end of this unit you will burn food and show that it gives off heat. In your body the food you eat is 'burnt' in a controlled way during the process of respiration. Respiration allows the energy in food to be used by a living organism to keep it alive: respiration also gives off heat but that is energy that 'gets away'.

Table 16.1 shows the different daily needs of fairly active people. Only the foods mentioned in Unit 15 are shown: there are many other important vitamins and inorganic ions.

The amount of energy we need each day depends on who we are and what we are doing. Table 16.1 has a column which shows a recommended daily energy intake for people of different ages and of both sexes and for pregnant and lactating women. (A lactating woman is one producing milk to feed a baby.) This table gives data for people who are fairly active; people who take little exercise will need less energy; people who are extremely active will need more.

The energy in food can be measured in **joules** (J). One gram (g) of butter, which is about what we eat in a mouthful of bread and butter, contains more than 30 000 J of energy. This gives you some idea how small 1 J is. To save writing lots of noughts, we usually give our energy needs in kilojoules (kJ or thousands of joules) or in megajoules (MJ or millions of joules). The recommended daily energy intake in Table 16.1 is given in MJ.

Table 16.1 Recommended daily intake

	Body mass in kg	Energy food (carbohydrate and lipid) in MJ	Protein in g	Vitamin C (ascorbic acid) in mg	Vitamin D in μg	Calcium in g	Iron in mg
Children							
under 1	7	3.4	14	20	10.0	0.5–0.6	5–10
1–3	13	5.7	16	20	10.0	0.4–0.5	5–10
4–6	20	7.6	20	20	10.0	0.4–0.5	5–10
7–9	28	9.2	25	20	2.5	0.4–0.5	5–10
Males							
10–12	37	10.9	30	20	2.5	0.6–0.7	5–10
13–15	51	12.1	37	30	2.5	0.6–0.7	9–18
16–19	63	12.8	38	30	2.5	0.5–0.6	5–9
adult	65	12.6	37	30	2.5	0.4–0.5	5–9
Females							
10–12	38	9.8	29	20	2.5	0.6–0.7	5–10
13–15	50	10.4	31	30	2.5	0.6–0.7	12–24
16–19	54	9.7	30	30	2.5	0.5–0.6	14–28
adult	55	9.2	29	30	2.5	0.4–0.5	14–28
Pregnant female (last 5 months)		10.7	38	30	10.0	1.0–1.2	14–28
Lactating female (first 6 months)		11.5	46	30	10.0	1.0–1.2	14–28

Source: *Handbook on Human Nutritional Requirements* (WHO, 1974).

Books on human diets often give the energy values in kJ of different foods. In Experiment 16.1 you will find the approximate energy value of a food by burning a known mass of it under a known volume of water and measuring the temperature rise of the water. Such measurements do not allow for heat that is given off into the air or elsewhere without heating the water. The apparatus used for official measurements of energy values of foods succeeds in reducing this heat loss but not in preventing it altogether: even official measurements are not completely accurate.

Table 16.2 shows how the energy used by a 65 kg man and 55 kg woman varies with how active they are. Questions at the end of this unit will give you practice in comparing different foods.

In 1977 *Dietary Goals for the United States* was published by a Senate Select Committee on *Nutrition and Human Needs*. Health workers had become alarmed because the usual American diet was rich: it was high in meat, lipid, refined foods and sugar. The Committee was particularly concerned because a high proportion of the lipid eaten was **saturated**. Saturated lipids tend to be those that are solid at room temperature, **unsaturated** those that are liquid at room temperature.

Saturated lipids are so called because they have the maximum possible number

Table 16.2 Energy used in different activities by a 65 kg man and a 55 kg woman

	Energy used over 24 hours in MJ					
	Little activity		Average activity		Great activity	
	man	woman	man	woman	man	woman
Asleep (8 hours)	2.1	1.8	2.1	1.8	2.1	1.8
At work (8 hours)	4.6	3.3	5.8	4.2	10.0	7.5
Domestic (8 hours)	4.6	3.2	4.6	3.2	4.6	3.2
Total	11.3	8.3	12.5	9.2	16.7	12.5

of hydrogen atoms in their fatty acids. They are thought more likely to damage your health than unsaturated lipids.

As well as eating rich foods, Americans were much less active than they had been. Not only in the United States but throughout the world a rich diet and lack of exercise go with heart disease, cancer, diabetes, tooth decay, overweight, high blood pressure and strokes.

No one suggests that all ill health is caused by diet and lack of exercise. The fact that certain diseases run in families tells us that inheritance may cause ill health. So may injury, smoking, drinking and drug-taking. Unit 18 describes diseases caused by smoking. We cannot change what we inherit and we may be unlucky enough to have accidents, but we choose what we eat and drink and whether we smoke or take drugs. Many people who know about the dangers of smoking and drug-taking do not realise how much they are harming themselves by what they eat.

Figure 16.5 summarises the American diet and the *Dietary Goals* recommended by the Senate Committee in 1977.

The Committee also recommended that people should reduce their salt intake to about 3 grams a day.

The exciting thing about the *Dietary Goals* is that they worked. Americans did change their eating habits and did take more exercise with the result that since 1977 there has been a marked fall in the United States in certain forms of ill health, notably heart disease.

(b) Heart disease
In the UK there have also been campaigns to get people to change their eating   193

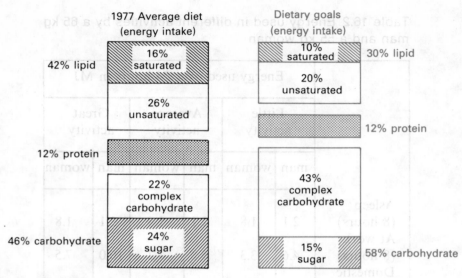

Figure 16.5   1977 diet and dietary goals

habits. They started after those in the United States and not so much has been spent on them. There has been only a slight fall in heart disease. Figure 16.6 shows that in the early 1980s the UK had one of the worst records for **coronary heart disease** (the cause of heart attacks) in the world. (Heart diseases are explained in greater detail in Section 19.4.)

A few years ago doctors at a medical school in Chicago persuaded people to come off the drugs they were taking for high blood pressure. They divided these people into two groups. The people in one group were given simple dietary advice, like that in *Dietary Goals*, and were asked to lose 4.5 kg in weight and to have

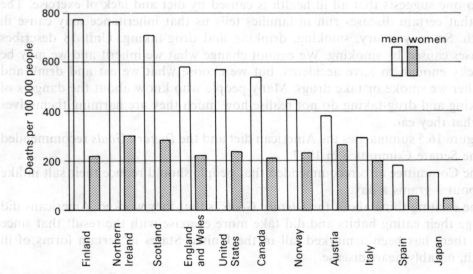

Figure 16.6   Death rates from coronary heart disease of people aged 35 to 70 in
194   eleven countries (source: WHO)

not more than two alcoholic drinks a day. The people in the other group were given no advice. Neither group was asked to give up smoking. Four years later 40 per cent of those given the dietary advice were continuing without drugs while almost all those in the other group had been back on drugs after a year. The other 60 per cent given the dietary advice were able to take fewer drugs.

(c) Cancer
There are also links between diet and certain cancers. Too little roughage (dietary fibre) is linked with cancer of the large intestine. Lack of roughage results in **constipation**, which means that undigested food takes longer to pass through the gut. Just why this should cause cancer is uncertain.

A study in China showed that a cancer of the oesophagus (the part of the gut that passes through the chest between the mouth and stomach) was related to diet. This cancer was common not only in the people of Linhsien county but also in the chickens that they kept around their homes. When a dam was to be built in Linhsien, a number of people were moved from Linhsien to Chungshan, where this form of cancer was rare. The Linhsien people did not take their chickens with them but bought new ones from flocks that were free of this gut cancer. Within five years a number of their chickens had developed the gut cancer, while the chickens of other people in Chungshan had stayed free of cancer.

Because the chickens were given table scraps in their food, a study was made of the Linhsien people's food to see if it could be giving both them and their chickens cancer. It was found that they made themselves a pickled mixture of vegetables which often developed a fungal growth after months in the pot. When analysed, this pickle was found to contain a high concentration of nitrosamines, which have long been thought to produce cancers. This is just one example of the link between diet and cancer.

(d) Obesity (overweight)
In 1983 the Royal College of Physicians published a report on *Obesity* and its relation to the death rates of men and women. They took as normal weight that which produced the lowest death rate in the records of life-insurance companies. But normal weight was altered to allow for people's height: taller people were expected to weigh more, shorter people less. **Obesity** was defined as 120 per cent or more of normal weight. The report also analysed the differences in the death rates of smokers and non-smokers. Figure 16.7 shows that both men and women are more likely to die young if they are overweight and if they smoke. Size on the horizontal axis takes account of both mass (what we call our weight) and height.

The report showed that overweight people were more likely to get heart disease, high blood pressure, diabetes, gall-bladder disease and gout.

(e) Additives
In recent years there has been much concern about **additives**. These are substances added to food for a number of reasons: to help manufacture it; to help preserve it; to improve its taste, texture and eye-appeal; even to increase its nutritional value, as when iron and calcium are added to flour. Additives have been used for 195

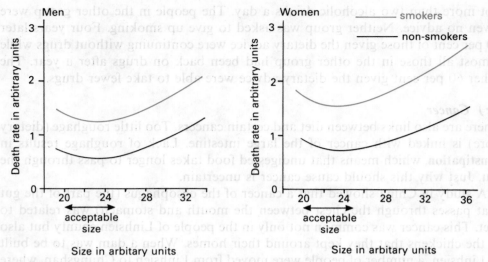

Figure 16.7   Association between body size, smoking and death rates for men and women (source: *Report of the Royal College of Physicians*, 1983)

many years. Section 15.5. deals with the addition of iron and calcium to flour in the UK.

Additives enable a greater variety of food to be provided and they help keep costs down by preventing food from going bad or having to be preserved by more expensive ways such as refrigeration. Many additives are harmless. It is the amount, as well as the nature, of the additives which is causing concern. In the early 1980s people in the UK were eating on average about 1.35 kg of additives a year: that is like taking 10 pills a day.

Caramel, not the sticky toffee sort made from sugar but that made by heating carbohydrates with ammonia and sulphur dioxide, is used for colouring, not flavouring. Used a thousand times more than any other colouring agent, caramel is suspected of causing genetic mutations (changes in inherited material) and cancer.

Monosodium glutamate, another common food additive, is described as a flavour-enhancer. It has no flavour of its own but, by stimulating taste buds in the mouth, brings out the flavour of substances, such as packet soups, to which it is added. It is banned from foods intended for babies or young children.

Additives are now listed as E numbers on many food packets but not many people know what the E numbers stand for. E numbers identify substances regarded as safe by all the Governments of the countries in the European Economic Community (EEC). Substances are removed from the list if they are thought to be unsafe. Caramel is E150. Monosodium glutamate does not, at the time of writing, have an E number. The reason for having the E numbers is that they are the same in all the different languages of the EEC.

16.5 Recommendations

196   So what does all this information on diet and lifestyle add up to? The advice is:

○ Limit the saturated (mostly animal) fats you eat
○ Eat a lot of roughage (dietary fibre) in whole cereal grains, vegetables and fruit
○ Limit the salt you eat
○ Keep your weight within 'normal' limits
○ Take exercise
○ Do not smoke

Questions

Q 16.1. Table 16.3 shows the food eaten in one day by a fairly active 55 kg woman.

Table 16.3 Food eaten in one day by a 55 kg woman

Meal	Food and drink	Quantity	Energy in kJ	Protein in g	Lipid in g	Carbohydrate in g	Calcium in mg	Iron in mg	Vitamin C in mg
Breakfast	white bread	90 g	950	7	2	50	90	1	0
	butter	15 g	450	0	12	0	2	0	0
	jam	30 g	330	0	0	19	5	0	1
	black coffee	1 cup	20	0	0	1	4	0	0
Lunch	hamburger	150 g	1 560	30	15	30	50	4	0
	ice cream	100 g	800	4	12	20	130	0	1
	fizzy drink	1 can	550	0	0	30	0	0	0
Evening meal	sausages	75 g	1 150	9	24	10	30	0.5	0
	potato chips	200 g	2 100	8	20	70	25	2	20
	baked beans	220 g	600	10	1	20	100	2.5	4
	apple pie	150 g	1 800	5	25	60	60	1	1
	cream	30 g	550	0.5	15	1	20	0	0
	tea with milk	2 cups	200	2	4	6	100	0	0
Snacks	chocolate	50 g	1 200	5	20	25	120	1	0
	peanuts	50 g	1 200	15	25	5	30	1	0
TOTAL INTAKE FOR DAY			13 460	95.5	175	347	766	13	27

(a) Compare her day's intake with that recommended in Table 16.1 and answer the following questions.
   (i) By how much was her energy intake greater than the recommended amount?
   (ii) What will be the likely effect if she continues to have an average daily energy intake of this amount for several weeks?
   (iii) Which substances did she not get enough of on that day?
(b) (i) Which food had the highest vitamin C content per gram? What was the mg per gram vitamin C content of this food?
   (ii) Which food had the highest carbohydrate content per gram?

197

(c) Choose *one* of this day's foods and suggest an alternative that would have been healthier. Give a reason for your answer.

Q 16.2. Table 16.4 shows food values in different pints of milk.

Table 16.4 Food values in a pint of milk

	Pasteurised	Homogenised	Channel Island	Semi-skimmed	Skimmed	Sterilised
Calcium	702 mg	702 mg	702 mg	729 mg	761 mg	702 mg
Energy	1.59 MJ	1.59 MJ	1.86 MJ	1.17 MJ	0.82 MJ	1.59 MJ
Total lipid including	22.2 g	22.2 g	28.1 g	10.5 g	0.6 g	22.2 g
saturated lipid	13.2 g	13.2 g	16.8 g	6.3 g	0.4 g	13.2 g
Protein	19.3 g	19.3 g	21.1 g	19.5 g	19.9 g	19.3 g
Carbohydrate (lactose)	27.5 g	27.5 g	27.5 g	28.4 g	29.3 g	27.5 g.

(a) Which milk has the highest energy content?

(b) Skimmed milk has the lowest energy content. Use information in the table to explain why.

(c) (i) Which *three* milks are shown in the table with exactly the same food values?

(ii) Explain the *differences* between these three milks.

Q 16.3. Table 16.5 shows the percentage of lipid, and the percentage of the lipid which is saturated, in a variety of foods.

Table 16.5 Lipid and saturated-lipid content of seven foods

Food	Milk	Butter	Soft margarine	Sunflower oil	Cheddar cheese	Roast lamb	Steamed plaice
% lipid content	4	82	80	100	34	26	2
% of lipid which is saturated	60	60	30	15	60	50	20

(a) Should you have more saturated or unsaturated lipid in your diet?

(b) Which of the foods in the table has the highest proportion of its lipid in an *unsaturated* form?

(c) Which of the following pairs contains less saturated lipid? (i) 100 g roast lamb or 100 g soft margarine? (ii) 100 g sunflower oil or 100 g soft margarine?
(d) Which *three* of the foods in the table have more of their lipid unsaturated than saturated?
(e) Which of milk, butter and cheddar cheese has the highest proportion of saturated lipid?
(f) Which of the foods in the table has the lowest proportion of saturated lipid?

Q 16.4. Table 16.6 shows the dry-mass content of 100 g of mycoprotein and 100 g of lean uncooked meat.

Table 16.6   Contents of mycoprotein and lean uncooked meat

	Dry mass in grams	
Contents	Mycoprotein	Lean uncooked meat
Protein	47	68
Lipid	14	30
Fibrous carbohydrate	25	0
Non-fibrous carbohydrate	10	0

(a) What is mycoprotein?
(b) Give reasons why mycoprotein might be a healthier food than meat.

Practical work

Experiment 16.1 To measure the energy content of food

Note to teachers
Seeds with a high lipid content burn easily and hold their shape as they burn. Castor-oil seeds and peanuts have high lipid contents and are large enough to be weighed individually. Dry any food to be used by leaving it in an incubator or oven at 100°C for several hours before the experiment begins.

> There is a risk that material may catch fire from burning foods in this experiment. Ensure that students have sufficient space to avoid harm to each other, that you can easily see all the students and that you have a fire blanket to hand.

Note to students
You have read in this Unit that energy is measured in units called joules (J).   199

Raising the temperature of 1 g of pure water by 1°C requires 4.2 J of energy. A volume of 1 cm^3 of pure water has a mass of 1 g. You will use this information in the experiment you are about to do.

Materials needed by each student
1 10 cm^3 graduated pipette fitted with a pipette filler
1 boiling tube
1 weighing bottle
1 glass rod
1 thermometer
1 pair of forceps
1 wooden-handled mounting needle
1 Bunsen burner
1 heat-resistant mat
1 clamp, stand and boss head
1 dried castor-oil seed

Materials needed by the class
balances to weigh to 0.05 g (1 per 4 students)
distilled water

Instructions to students
1. Before you start work, read through these instructions and make a note of what you are asked to record. Design a table in which you can record your results. Appendix A helps you design a table.
2. Put the clamp, stand and boss head together so that they hold the boiling tube vertically about 30 cm above the heat-resistant mat on the bench.
3. Use the pipette to put exactly 10 cm^3 of distilled water into the boiling tube. Use the thermometer to measure the temperature of this water. Record this temperature and replace the thermometer with the glass rod.
4. Weigh a seed and record its mass.
5. Stick the needle through the seed so that it holds the seed firmly. Holding the needle by its wooden handle, put the seed in a blue Bunsen flame until it burns with a bright yellow flame.
6. Hold the burning seed under the boiling tube of water. Carefully stir the water in the boiling tube with the glass rod. (Do not drop the rod on the bottom of the tube or the tube may break.) If the seed stops burning, put it back in the Bunsen flame to relight it.
7. When the seed will no longer burn, put it on the heat-resistant mat, stir the water in the boiling tube and measure its temperature with the thermometer. Record this temperature.
8. Use the forceps to crush the blackened remains of the seed. Record what happens.

Interpretation of results
1. Calculate the rise in temperature of the water.
2. To raise the temperature of 1 cm^3 of water by 1°C needs 4.2 J of energy. Calculate the amount of energy needed to cause the rise in temperature of the

water. To get the answer in joules you must multiply 4.2 by 10 (because you used 10 cm^3 of water) and then by the rise in temperature.

3. Finally, calculate the amount of energy which would have been released by burning 1 g of the seed. To do this you must divide the amount of energy released by the burning seed by the mass of the seed in grams.

4. If you can get a book with a table of food values, compare your result with the appropriate figure in the table.

5. Had *all* the stored food in the seed been burned? How can you tell?

6. Was *all* the energy released from the burning seed used to heat up the water? How do you know?

7. You can adapt this method to find the energy content of other dried foods.

MOVEMENT, SUPPORT AND EXERCISE

17.1 Movement

Muscles provide the power for us to move. Muscles are attached at various points throughout our body to the bones that form our skeleton. These muscles are our **skeletal muscles**; we can control them consciously and make deliberate fine movements with them. The control we have over skeletal muscles is different from the unconscious control we have over the muscles of the gut that bring about peristalsis (see Section 16.1).

Skeletal muscles, like the muscles of the gut, work **antagonistically**. There are always two sets: while one set contracts and shortens, the opposite antagonistic set relaxes and is lengthened by the pull of the contracting set. These actions move the bones and can be reversed: when the set that first contracted relaxes and the set that first relaxed contracts, the bones are moved back to their starting position. The power for the work done in moving the bones is provided by the contracting set of muscles. The energy needed for the muscles to contract comes from respiration.

Take a look at movement of the knee joint in Figure 17.1 to see antagonistic muscles. The muscles that bend and straighten the knee lie at the back and front of the thigh, alongside the **femur** bone. At one end they are joined to the **pelvis** and to the top of the femur; at the other end they are joined to the **tibia** (the shin

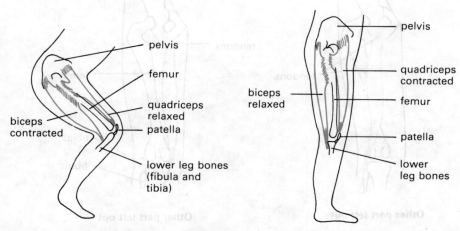

Figure 17.1   Knee bent and straightened

bone), to the **fibula** (a thinner bone behind the tibia) and to the **patella** (the small bone which forms the knee cap and which is firmly joined by **ligaments** to the tibia). The big muscle at the front is the **quadriceps** (made up of four parts); the muscle at the back is the **biceps** (made up of two parts). To bend the knee, the biceps contracts. To straighten the knee, the quadriceps contracts.

The front view of the thigh in Figure 17.2 shows the position and attachment of the four parts of the quadriceps. The back view of the thigh in Figure 17.3 shows the position and attachment of the two parts of the biceps.

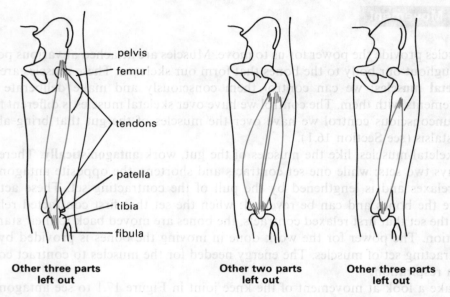

Figure 17.2   Front of thigh showing the four parts of the quadriceps

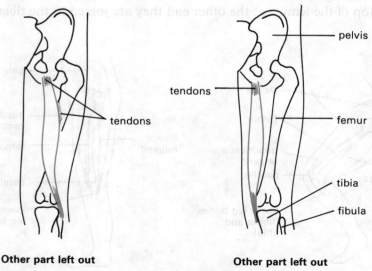

204   Figure 17.3   Back of thigh showing the two parts of the biceps

To get an idea of how these muscles work, stand up straight and, by raising one leg slightly off the ground, bend it slowly and completely at the knee while keeping your body straight. Feel the biceps muscle contract and harden at the back of your thigh. Meanwhile the quadriceps at the front has had to relax and lengthen. If you straighten your leg again slowly, you will feel the quadriceps contract and harden at the front of your thigh while the biceps at the back relaxes and is lengthened.

The muscles around the thigh are wide but in front of and behind the knee they narrow and merge into tough **tendons** by which they are joined to the lower-leg bones: you can feel the two tendons of the biceps behind your knee when you contract the biceps muscle. You can feel a wider tendon that joins one of the quadriceps parts to the tibia just below your knee cap and yet another tendon just above your knee cap. The tendons are shown in Figures 17.2 and 17.3.

Tendons are tough and do not stretch: when the muscle contracts, the pull is transferred to the bone and is not wasted in stretching the tendon. **Ligaments**, on the other hand, join the bones to one another and must be able to stretch a bit when the joint is moved. When the knee is bent, the ligaments holding the bones together at the front must stretch a bit on the outside of the joint. When the knee is straightened, the ligaments must shrink again to hold the bones firmly.

○ Tendons join muscle to bone and are inelastic
○ Ligaments join bone to bone and are elastic

Both ligaments and tendons are white and tough, consist of **connective tissue** and contain few blood vessels. Torn ligaments and tendons are well known sports injuries: they take time to heal because they are short of blood vessels bringing food materials for repair.

(a) Muscle structure

A muscle, such as one of the four parts of the quadriceps, is shown cut across in Figure 17.4. It consists of thousands of fibres which run lengthwise through the muscle and are bound together in bundles of about a thousand. Surrounding each bundle and surrounding the whole muscle is white connective tissue which gets thicker and tougher to form the tendons where the muscle joins the bones.

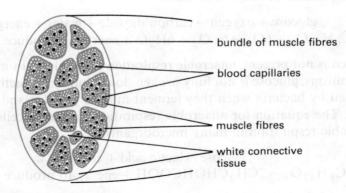

Figure 17.4   Section across muscle

You can see both long and short tendons joining the biceps and quadriceps to the bones in Figures 17.2 and 17.3.

The small red dots shown in the bundles between the muscle fibres are fine blood vessels called **capillaries**. They bring glucose and oxygen to the muscle fibres for aerobic respiration: the energy released from glucose in respiration enables the muscles to work. Intensive training can almost double the number of these capillaries and greatly improve athletic performance.

(i) Muscle contraction
Skeletal muscles are under the control of our will. When we want to move a bone, electrical signals in the form of a series of **nerve impulses** are sent through **nerve fibres** from the **brain** to the muscles. All the impulses are of the same strength; it is the time interval between the impulses that changes the message to the muscles.

If a muscle is stimulated by a single impulse, it contracts only in a quick small twitch which takes about 0.1 of a second; it then relaxes, which takes about 0.2 of a second. If a second impulse arrives in less than 0.2 of a second, the muscle will contract again before it has finished relaxing after the first contraction. It will therefore contract more. If further impulses arrive quickly enough, the point will be reached where the muscle is fully contracted. When impulses stop arriving, the relaxation of the muscle is no longer prevented. Provided no more nerve impulses arrive, the muscle will become completely relaxed. It is stretched to its greatest length when the antagonistic muscle contracts.

A single muscle fibre is a few cells wide but many cells long: in fact the cells are merged together and have a complex structure enabling each fibre to contract as a whole. It is respiration in the fibres that releases energy from glucose in the form of ATP; the ATP makes it possible for each fibre to contract when it is stimulated by a nerve impulse.

(ii) Energy supply
Unit 5 explains that respiration in the cells of microorganisms can take place either with oxygen or without it. The same is true of respiration in the cells of the muscle fibres. As in microorganisms, glucose contains the energy which, when oxygen is present, breaks down to carbon dioxide and water and releases energy as ATP. The equation for **aerobic respiration** in muscle cells is the same as for aerobic respiration in microorganisms:

$$\text{glucose} + \text{oxygen} \rightarrow \text{carbon dioxide} + \text{water} + \text{energy}$$
$$C_6H_{12}O_6 + 6O_2 \rightarrow 6CO_2 + 6H_2O + \text{energy to produce 38ATP}$$

If oxygen is not present, **anaerobic respiration** takes place. In muscle cells as in microorganisms, glucose is not fully broken down and forms **lactic acid** (the same acid formed by bacteria when they ferment milk to yoghurt and cheese or grass to silage). The equation for anaerobic respiration in muscle cells is the same as for anaerobic respiration in many microorganisms:

$$\text{glucose} \rightarrow \text{lactic acid} + \text{energy}$$
$$C_6H_{12}O_6 \rightarrow 2CH_3CHOHCOOH + \text{energy to produce 2ATP}$$

206   At rest and during normal activity there is enough oxygen in the muscle cells

for aerobic respiration to take place. During extreme activity, such as sprinting 100 metres, the oxygen in the muscles is used up and the blood cannot bring more quickly enough. Anaerobic respiration then takes place. If a lot of lactic acid is formed in the muscle, it is no longer able to continue contracting: this tiredness or **fatigue** must be followed by a period of rest or **recovery**.

During recovery from the effects of anaerobic respiration about one-fifth of the lactic acid that was produced is broken down with the use of oxygen in aerobic respiration to release more energy. This energy is used to reconvert the remaining four-fifths of the lactic acid back to glycogen. It is said that this oxygen is needed to repay the **oxygen debt**.

(iii) Energy reserves
Energy for the work of skeletal muscles is stored in the carbohydrates and lipids described in Sections 15.2 and 15.3. Most of the energy reserves are in the form of the lipid **triglyceride** under the skin and around certain organs; some is in the form of the carbohydrate **glycogen** in the liver and in the muscles; a very small amount circulates in the blood as **glucose**. One gram of triglyceride releases 35 kJ of energy when it is burned; one gram of glycogen releases only 16 kJ. Lipid is therefore a more efficient energy reserve than glycogen.

Triglyceride is reconverted to fatty acids and glycerol before the energy in it can be released in respiration. The fatty acids do not have to be converted into glucose before the energy in them can be released in respiration. During respiration in the cytoplasm of the muscles, the fatty acids are used directly to release ATP. Figure 17.5 shows how the three energy reserves (triglycerides, glycogen and glucose) can all be used by a muscle.

Triglycerides are plentiful, but it takes time to convert them to fatty acids and transport them to muscles. Liver glycogen is not plentiful and is needed as a reserve for all the cells of the body, not only for muscle cells, but it can be converted more quickly than triglycerides; muscle glycogen is the scarcest of all, but it is in the right place. Table 17.1 shows how much of these three energy reserves and of blood glucose is available in an athlete for work by skeletal muscles and how long each would last in continuous walking and in running a marathon.

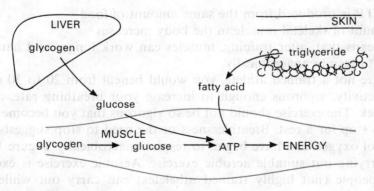

Figure 17.5   Use of energy reserves in a muscle

207

Table 17.1 Energy reserves for skeletal muscles in an athlete

Energy reserve	Mass in grams	Energy in kJ	Time the reserve would last in minutes	
			Walking	Marathon running
Triglyceride	9 000	337 500	15 500	4 018
Liver glycogen	100	1 660	86	20
Muscle glycogen	350	5 800	288	71
Blood glucose	3	48	2	< 1

17.2 Exercise

There is no doubt that regular exercise improves health. Studies in the United States have shown that regular exercise can add up to 10 years to a person's life. Section 18.4 tells how regular exercise improves breathing and Section 19.4 how it improves the working of the heart. In the studies in the United States a group of elderly women who took part in training sessions three times a week over a period of three years did not lose any calcium from their bones while a control group, who did not train, did lose calcium (see Section 15.5).

Regular vigorous exercise, such as that taken by someone training for a marathon, can produce such dramatic effects as doubling the number of blood capillaries between the muscle fibres. Much less vigorous exercise can bring a worthwhile increase in the number of these blood capillaries. The result is an increase in the amount of oxygen and glucose that the blood brings to the muscles.

Regular exercise improves energy release in the muscles in other ways too:
o the muscle fibres become better at extracting oxygen and glucose from the blood in the capillaries
o the concentration of enzymes in the muscle fibres concerned with respiration increases
o more ATP is produced from the same amount of food
o the amount of skeletal muscle in the body increases
All this means that, after training, muscles can work longer and harder before they start to respire anaerobically.

If you are not a trained athlete, you would benefit from 20 to 30 minutes of physical activity, vigorous enough to increase your breathing rate, about four times a week. The exercise should not be so vigorous that you become breathless or have to stop for a rest. Breathlessness or the need to stop suggests you have run short of oxygen and have begun to respire anaerobically. Figure 17.6 shows a class carrying out suitable aerobic exercise. Aerobic exercise is exercise that ordinary people (not highly trained atheletes) can carry out while respiring aerobically.

Figure 17.6   Aerobic exercise to keep you fit

17.3 Support

The skeleton supports the body and gives it shape. Unfortunately, the **vertebral column** (or backbone) causes a lot of trouble in humans. This is because it evolved in four-legged animals and is less suited to two-legged ones. The vertebral column is made up of strong but separate bones called **vertebrae**. The fact that the vertebrae are separate allows the vertebral column to bend a little. Between each vertebra and the next is a shock-absorber, an **intervertebral disc**, held in place by ligaments. A normal vertebral column is curved at the base of the neck and at the back of the abdomen as shown in the side view in Figure 17.7. You can see the vertebrae and the discs in Figure 17.7. What you cannot see is that the discs are of different sizes and shapes which fit the spaces between the vertebrae exactly.

You have probably heard of a 'slipped disc'. Provided that you stand and sit correctly, the pressure on the edges of the discs in your vertebral column is even. If you stand or sit for long periods, so that the natural curves in the vertebral

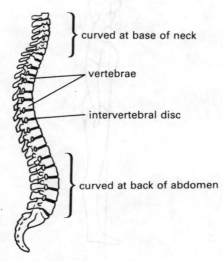

Figure 17.7   Curved vertebral column

209

column are either increased or straightened out (or even bent the wrong way), the weight on a part of the edge of the disc is unnaturally increased and the disc may crack or it may break through the ligaments surrounding it (or it may do both). **A slipped disc** is one that has broken through the surrounding ligaments. Standing and sitting incorrectly also place a strain on the muscles, which are forced to hold the vertebrae in an unnatural position.

(a) Standing

The way to stand correctly, which means the way to put even pressure on the edges of your discs, is to
○ lift your chest away from your abdomen and relax your shoulders
○ make yourself as tall as possible but keep your head relaxed and your chin parallel to the ground
○ pull your abdomen in (a potbelly increases the abdominal curve of the vertebral column)
○ put the weight of your body forward on the balls of your feet (not back on your heels)
○ keep your feet slightly apart at the heels and slightly more apart at the toes
A correct standing position is shown in Figure 17.8.

(b) Sitting

The way to sit correctly, which again means the way to put even pressure on the edges of your discs, is to
○ lift your chest away from your abdomen and relax your shoulders
○ pull your abdomen in
○ bend your knees at an angle of about 90°

You can see from Figure 17.9 that bending at the hips at 90° decreases the normal curve of the vertebral column in the abdominal region. (Sitting on the ground with your legs straight out in front of you decreases it even more.) Sitting

210   Figure 17.8   Correct standing

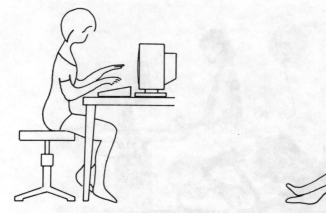

Figure 17.9   Sitting without a back support   Figure 17.10   Dangerous sitting position

with your back unsupported is all right for a short time, but it puts a strain on both the discs and the muscles. If you sit for long on an ordinary chair, your back should be supported to keep it curved forward (and not straight or curved backwards).

Figure 17.10 shows the worst sort of sitting position with the back unsupported and curved the wrong way. Sitting like this opens up the spaces at the back of the vertebrae and compresses the discs at the front: the result may be that a disc slips out at the back.

The fact is that even 'sitting correctly' on an ordinary chair is unnatural. Primitive people do not sit, and this is not only because they have no chairs. They do not sit on rocks of a convenient height as we would do. When they do not want to stand or lie, they squat. Squatting keeps the vertebral column in its natural curves.

Figure 17.11 shows chairs that encourage a sitting position closer to squatting. The curve of the vertebral column is kept without muscle strain. The weight of the body does not rest only on the base of the vertebral column but is partly taken up by the knees. The vertebral column is kept in its natural curves, which is why it does not need supporting. Such chairs are now widely used, particularly by VDU operators and typists, who in the past suffered a lot from backache.

(c) Lifting and carrying

Figure 17.12 shows the right and the wrong ways to lift a heavy weight. You can see that the wrong way has the vertebral column curved in the wrong direction: there is a danger that a disc compressed at the front will slip out of place backwards and there is a strain on the muscles. Both the danger of the slipped disc and the strain on the muscles are increased by the heavy weight and by the fact that the heavy weight is some distance in front of the legs.

The correct way to lift a heavy weight (you should avoid lifting very heavy ones) is to

○ put your feet either side of the weight and as near to the weight as possible
○ bend your knees
○ keep your back as straight as possible

211

Figure 17.11   Back chairs

Figure 17.12   Right and wrong ways to lift

○ lift by straightening your knees
○ lift slowly (do not jerk)

The correct way to lift something heavy, such as a suitcase, with one hand is to

○ put your feet as close to it as possible
○ keep one foot a little in front of the other
○ bend your knees
○ keep your back as straight as possible
○ lift by straightening your knees
○ lift slowly (do not jerk)

When you carry a heavy weight in one hand, do not let it pull you down on one side but lean in the opposite direction. Carrying a heavy suitcase in one hand is something you should be able to avoid. It is better to pack your things in two small suitcases and carry one in each hand.

Questions

Q17.1

(a) Make a table like the one shown below but larger. Fill in the details for tendons and ligaments.

	Appearance	Parts joined	Elastic or inelastic
Tendons Ligaments			

(b) Explain why torn tendons and ligaments heal more slowly than torn muscles.

Q17.2

(a) What are the energy sources used by a runner in a marathon? Where do they come from?

(b) In what ways is energy release in the muscles different in the 100 m sprinter and the marathon runner?

(c) What is meant by an 'oxygen debt'? What happens when it is repaid?

Q17.3

(a) Why is it important to keep the natural curves of the vertebral column even when sitting?

(b) Look at the people sitting on the back chair in Figure 17.11. How does this chair help to prevent backache?

AIR SUPPLY

18.1 Introduction

Air contains 20 per cent oxygen. We need oxygen to release energy in aerobic respiration in our cells. We get oxygen from air. We draw air into our bodies when we breathe; Figure 18.1 shows the series of tubes passing from the nose and mouth through the neck and down into the **lungs**, where oxygen in the air is passed into the blood. Contraction and relaxation of the chest muscles surrounding the lungs result in breathing or **ventilation**. After air has been breathed into the lungs, air with less oxygen but more carbon dioxide (and water vapour) is breathed out from the lungs.

Lungs consist of millions of **air sacs** (**alveoli**) and narrow tubes (**bronchioles**) that bring air to them and take air away from them. The air sacs are formed like bunches of grapes at the ends of the narrow tubes, which branch out everywhere through the lungs. Outside the air sacs, and touching them, is a network of fine blood **capillaries**. All this is shown in Figure 18.2.

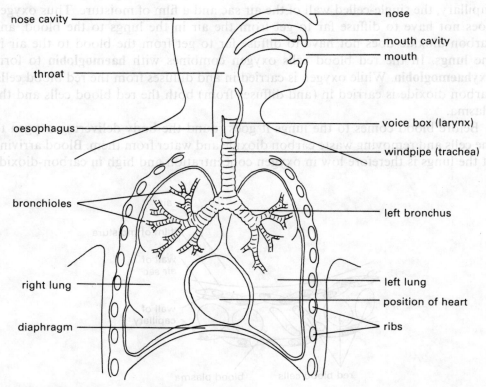

Figure 18.1   Ventilation (breathing) system

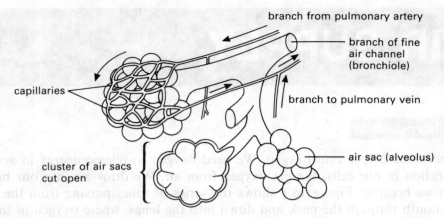

branch from pulmonary artery

branch of fine
air channel
(bronchiole)

capillaries

branch to pulmonary vein

air sac (alveolus)

cluster of air sacs
cut open

Figure 18.2   Air sacs and capillaries

18.2 Exchange of gases

The inner surfaces of the air sacs provide a huge moist surface area for the diffusion
of oxygen into the blood and the diffusion of carbon dioxide (and water) from
the blood. Blood passing through the capillaries is only a short distance from air
in the air sacs of the lungs. Figure 18.3 shows how a red blood cell is separated
from air in an air sac only by **plasma** (blood fluid), the single-celled wall of the
capillary, the single-celled wall of the air sac and a film of moisture. Thus oxygen
does not have to diffuse far to get from the air in the lungs to the blood, and
carbon dioxide does not have to diffuse far to get from the blood to the air in
the lungs. In the red blood cells oxygen combines with **haemoglobin** to form
oxyhaemoglobin. While oxygen is carried in and diffuses from the red blood cells,
carbon dioxide is carried in (and diffuses from) both the red blood cells and the
plasma.

Before blood comes to the lungs it goes round the body delivering oxygen to
the cells and removing waste carbon dioxide and water from them. Blood arriving
at the lungs is therefore low in oxygen concentration and high in carbon-dioxide

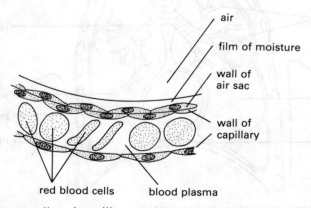

air

film of moisture

wall of
air sac

wall of
capillary

red blood cells   blood plasma

Figure 18.3   Air-sac wall and capillary

concentration. Water evaporates from the moist air-sac lining (see Section 20.5), but controlled removal of excess water from the blood takes place only in the kidneys.

All the essentials for efficient gas exchange by diffusion are present:

○ there is a large surface area (the lining of the air sacs)
○ there is a short diffusion distance (between air in the air sacs and blood)
○ there are steep diffusion gradients (between high oxygen concentration in air and low oxygen concentration in the red blood cells and between high carbon-dioxide concentration in blood and low carbon-dioxide concentration in air)

18.3 Variations in depth and rate of ventilation (breathing)

The lungs of a human adult hold about 6 dm^3 of air. When we are resting and ventilating (breathing) normally, about 0.5 dm^3 of air passes in and out of our lungs about fifteen times a minute (0.5 dm^3 × 15 = 7.5 dm^3 of air per minute). This is called **tidal air**. As soon as we become more active, the muscles use up more oxygen and produce more carbon dioxide. When the lower oxygen or the higher carbon-dioxide content of the blood is detected by the body, stronger and quicker contractions of the chest muscles are started. The air moves into and out of the lungs both in greater amount and more quickly. As a result more oxygen is absorbed per minute by the red blood cells and more carbon dioxide is removed per minute from the blood.

Not all the air in the lungs is breathed out; the most we can breathe in or out is about 4.5 dm^3. Figure 18.4 shows approximate lung volumes:

total volume	6.0 dm^3
tidal volume	0.5 dm^3
maximum breath volume (**vital capacity**)	4.5 dm^3
volume always remaining in lungs after breathing out as strongly as possible (**residual volume**)	1.5 dm^3

The student in Figure 18.5 is using a piece of apparatus called a **recording spirometer**. From the trace it makes on a revolving drum, the volume of air breathed in and out and the time taken for each breath can be calculated.

The principle on which the spirometer works is shown in Figure 18.6. The person breathes in and out of the tube through the mouth. As air is breathed in, the volume in the chamber falls and the floating lid sinks: the marker pen falls while a recording drum turns at a constant rate and the pen makes a diagonal mark on the drum paper. As air is breathed out, the volume in the chamber increases and the floating lid rises: the marker pen rises while the recording drum continues to turn at a constant rate and again the pen makes a diagonal mark on the paper. The paper on the drum can be calibrated so that vertical distances on the paper show the volumes of gas breathed in and out. Given the speed at which the rotating drum turns, the number of breaths per minute can be worked out from the line on the paper.

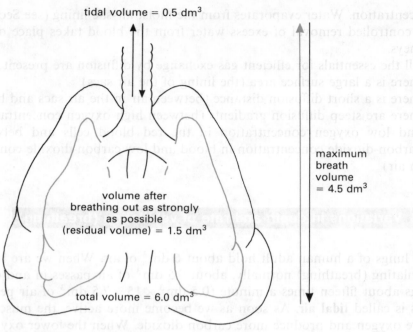

tidal volume = 0.5 dm³

maximum breath volume = 4.5 dm³

volume after breathing out as strongly as possible (residual volume) = 1.5 dm³

total volume = 6.0 dm³

Figure 18.4   Lung volumes

It would be dangerous for you to use a spirometer without a teacher present: if you used up the oxygen in the chamber and carried on breathing through it, the oxygen in your blood would fall: you could stop breathing and could even die.

In the practical work at the end of this Unit you will use a simpler method to measure the rate of breathing (that is, the number of breaths per minute).

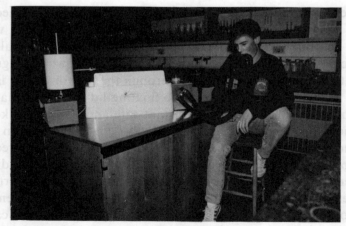

Figure 18.5   Student using a recording spirometer

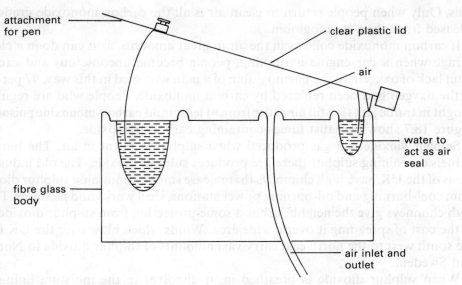

attachment for pen

clear plastic lid

air

water to act as air seal

fibre glass body

air inlet and outlet

Figure 18.6   Spirometer

Section 17.2 explains that regular exercise improves the efficiency of muscles and helps old people to keep the calcium in their bones. It can also improve the efficiency of gas exchange at the air-sac surface by increasing the volume of air drawn into the lungs and the number of breaths that can be taken (without stress) per minute.

At rest less than 1 per cent of the energy we are using goes on the muscular effort of breathing: we are not even aware that we are breathing. In a long hard race 10 per cent of the energy used by a fit athlete goes on breathing.

Deep and rapid breathing *after* a 100-metre race is necessary to pay off the **oxygen debt** that has developed during the race (see Section 17.1). The extra oxygen, that over and above normal needs, is used to remove the lactic acid produced during anaerobic respiration of the muscles (see Section 17.1).

18.5 Polluted air

The air we breathe in contains harmful things. Harmful bacteria and viruses that enter our bodies in air are usually made harmless by our white blood cells. More serious are harmful substances in air which come from our industrial society.

Carbon monoxide is a gas produced when petrol or coal is not completely burnt up to form carbon dioxide. It is also part of the smoke inhaled by a smoker. When carbon monoxide is breathed in, it combines with haemoglobin in the red blood cells that pass the air sacs. It therefore takes up space that oxygen should occupy and reduces the amount of oxygen that can be carried by the red blood   219 cells. Only when people return to clean air is all the carbon monoxide gradually released from their haemoglobin.

If carbon monoxide collects in the air in great amounts, as it can do in a closed garage when a car engine is running, people become unconscious and can die from lack of oxygen. In the haemoglobin of a man who died in this way, 97 per cent of the oxygen had been replaced by carbon monoxide. People who are regularly caught in traffic jams in still air suffer from at least mild **carbon-monoxide poisoning**. Figure 18.7 shows exhaust fumes containing carbon monoxide.

Sulphur dioxide is a gas produced when sulphur is burnt in air. The burning of fuels containing sulphur therefore produces sulphur dioxide. The old industrial areas of the UK have high chimneys that release smoke containing sulphur dioxide from coal-burning and oil-burning power stations, steel works and factories. These high chimneys give the neighbourhood some protection from sulphur dioxide but at the cost of spreading it over a wide area. Winds which blow over the UK from the south west to the north east carry vast amounts of sulphur dioxide to Norway and Sweden.

When sulphur dioxide is breathed in, it dissolves in the moisture lining the lungs and forms an acid. The acid irritates and may damage the delicate air-sac surfaces, increasing the amount of mucus produced. Delicate, hair-like **cilia** on the surface of the air passages in the lungs and throat normally waft mucus into the back of the throat, where it is swallowed without being noticed. When a lot of sulphur dioxide is breathed in, the cilia are unable to cope with the increased mucus and stop working. The mucus stays in the lungs. People cough to bring up excess mucus **(phlegm)** from their lungs.

220 Figure 18.7 Exhaust fumes containing carbon monoxide

Constant coughing damages the thin dividing walls between the clusters of air sacs with the result that the area for gas exchange is reduced. Viruses and bacteria infect the damaged lungs and lead to fever. Scar tissue forms in the lungs and the fine air channels leading to the air sacs get narrower. All this means that greater effort is needed to get enough oxygen into the blood.

People to whom this has happened are suffering from **bronchitis**, so called because of the irritation in the fine air channels or **bronchioles**. In many countries bronchitis is known as the 'English disease' because it is so common in England. (Bronchitis is the new English disease; the old one was rickets.) For people with chronic (long-lasting) bronchitis, breathing is a continual effort using up a lot of their energy.

The damaging effect of sulphur dioxide was brought home to people in the UK by a severe fog in London in 1952. The fog was called 'smog' because of the amount of smoke in it. Figure 18.8 shows the amount of sulphur dioxide in the air before, during and after the fog and the number of deaths recorded each day. As you can see, more than three times as many people died on days during and immediately after the fog as on days before it. Many of those who died were already ill with bronchitis or heart problems, but they died sooner than they would have done had there been no fog. These startling figures led to a number of Clean Air Acts, which have improved conditions in most cities by, for example, making people use smokeless fuel. The amount of sulphur dioxide released into the air has nonetheless gone up, not down. It has merely been spread wider.

Dust from coal and other industrial materials has much the same effect upon the lungs as sulphur-dioxide gas. When breathed in, it causes irritation resulting in excess mucus that cannot be got rid of. **Pneumoconiosis** is the name given to lung diseases caused by breathing in dust in air. There are various forms of pneumoconiosis: coal dust results in **anthracosis**, silica dust (from mining and sand-blasting) in **silicosis** and asbestos dust in **asbestosis**. The damage caused by

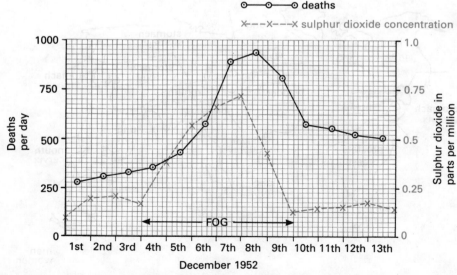

Figure 18.8   Deaths and sulphur dioxide in the air during a London fog

dust is now better understood and action is taken to prevent it. Unfortunately damage that has been done to the lungs cannot, at present, be repaired. The symptoms of the different forms of pneumoconiosis are the same as those of bronchitis: coughing and bringing up mucus.

Asbestos dust causes not only bronchitis but **lung cancer** and even cancer of the stomach and large intestine. Tiny fibres of asbestos lodge in the air sacs and may start the coughing that leads to bronchitis. They may also start uncontrolled division of cells in lung tissue and thus form a **tumour** which takes up more and more space and prevents the normal lung tissue carrying out its function of gaseous exchange.

The manufacturing of asbestos is much less dangerous than it used to be. People who are likely to disturb asbestos during their work, such as workmen who have to demolish old buildings that were insulated with asbestos, have to have a Health and Safety Executive licence to show they have been trained to handle it correctly, and they have regular health checks.

Figure 18.9 gives a very rough idea of the most common cancers that cause the deaths of men and women throughout the world. In men, lung cancer is the biggest killer in the industrial parts of the world but other cancers are bigger killers elsewhere. In women, breast cancer and cancer of the cervix are the main killers.

18.6 Smoking

We are constantly warned against smoking. What makes it so harmful?

Tobacco smoke contains hundreds of different chemicals. As the manufacturers discovered harmful substances, they tried to remove them. Filter-tipped cigarettes

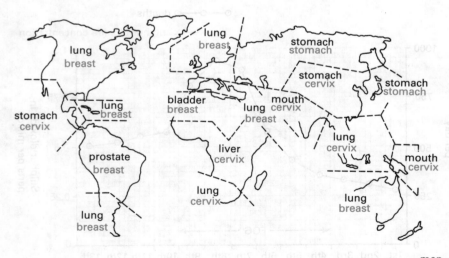

222   Figure 18.9 Biggest cancer killers of men and women world-wide

are intended to remove those that remain. But cigarette smoke still contains plenty of substances that cause ill health.

Like sulphur dioxide and dust, smoke irritates the inner lung surface causing increased mucus production and a slowing down and eventual stopping of the wafting movement of the cilia that line the breathing passages. Mucus collects in the lungs, causing 'smoker's cough' and eventually chronic bronchitis. The damage to health from bronchitis is the same whether it is caused by sulphur dioxide, dust or smoking. Figure 18.10 shows how much more common a cough and shortness of breath are among smokers than among non-smokers. It has been estimated that a fifty-year-old smoker has the lungs of a seventy-year-old non-smoker.

Nicotine is a compound in tobacco smoke which has a number of effects on the body. It acts on nerve cells, especially those of the brain: in small doses it increases brain activity yet in large doses reduces it. Nicotine releases hormones, especially adrenaline, into the blood; this increases the heart rate (or pulse rate), increases blood pressure and narrows blood vessels. The narrowing of blood vessels can lead, after many years, to damaged circulation, especially in the legs, which in extreme cases may have to be amputated. Nicotine may also increase the deposit of fatty substances lining the arteries, which increases the chances of blood-clotting. There is more about these effects in Unit 19.

Carbon monoxide is also present in smoke. As has been described above, carbon monoxide reduces the oxygen carried by the red blood cells. If women smoke when they are pregnant, the oxygen available for the **fetus** (the baby in the womb) is reduced, less energy is released by the fetus to make its cells and tissues, and it does not grow so well. We know this from the lower average birth weight of babies whose mothers have smoked during pregnancy.

Tobacco smoke causes damage to DNA (the important substance in the nucleus of cells) in smokers and in the placentas of pregnant women. Since damage to DNA is one of the causes of cancer, it has been suggested that this is why smokers are more likely to get lung cancer and other cancers too. The relation between smoking and cancer is, however, something into which more research is needed.

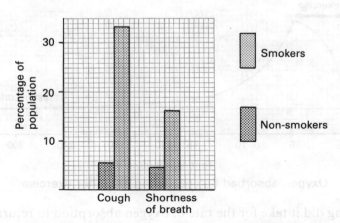

Figure 18.10   Cough and shortness of breath in the UK population

Non-smokers have always objected to being forced to breathe in tobacco smoke in enclosed areas and to getting the smell of stale tobacco smoke into their clothes and hair. Many non-smokers find that tobacco smoke makes them cough and sneeze and gives them sore eyes, a runny nose and a headache. People with contact lenses may find that smoke affects their sight and leads to eye infections. Recent studies have suggested that breathing in other people's tobacco smoke (known as **passive smoking**) can cause bronchitis and may make people more likely to get meningitis. In the UK smoking has long been forbidden on the lower decks of buses and in some of the carriages of trains. Non-smoking areas are widespread in public places: the London Underground is a total non-smoking area. Many older people are giving up smoking, but many young people are starting.

Questions

Q 18.1. Look at Figure 18.8.
(a) (i) On what day was the death rate highest? (ii) On what day was the increase in sulphur dioxide greatest?
(b) What does the graph show about the relation between deaths and sulphur dioxide in the air during the London fog?

Q 18.2. A student who was sitting down resting breathed in and out 450 cm^3 of air every 6 seconds.
(a) (i) How many breaths did she take each minute? (ii) How much air did she breathe in and out each minute?
(b) The air she breathed out was different from the air she breathed in: 4 per cent of the air she breathed in was not breathed out. (i) What was this gas? (ii) How much of this gas did she take from the air in each minute?
(c) What changes in her breaths per minute would you expect if she got up and ran away as fast as she could? Give reasons for your answer.

Q 18.3. Figure 18.11 shows the rate of oxygen absorption before, during and after exercise.

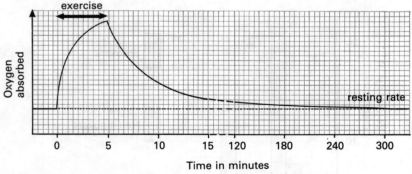

Figure 18.11 Oxygen absorbed before, during and after exercise

(a) How long did it take for the rate of oxygen absorption to return to the resting rate after the exercise?

224

(b) Why was extra oxygen needed *after* the exercise?

(c) Why was the graph drawn with a gap in it?

Experiment 18.1 To test the hypothesis that exercise affects the rate of human breathing

Note to teachers

> **Do not let students do this Experiment if they have medical problems which make it unsafe for them to exercise. Subject to this proviso, the Experiment can be done safely without supervision.**

You may decide that some students, particularly part-time students, should do it at home. The 'Instructions to students' assume that each of them will design the experiment and hand in an outline of the method and a list of practical requirements. If students' results are to be compared, time must be allowed to standardise the method.

Instructions to students

1. Design an experiment to test the hypothesis that exercise affects the rate of your own breathing. Some clues to help you design your experiment are given below. If after reading them and thinking about them you do not know how to start, look at Experiment 19.1. If you still do not know how to start, ask your teacher for help. You can do your experiment either by yourself or with someone to help you.

2. Make a list of the steps you will take during your experiment. Show it to your teacher, who will check that your method words.

3. The only equipment you will need is a watch or clock with a second hand. Tell your teacher if you do not have one.

4. When you have done your experiment, write a report telling what you did, what the results were and what these show about the hypothesis you are testing. Appendix B helps you write a report.

Clues to help you design your experiment

1. There are several ways of testing the hypothesis. You need not worry if your way is different from that of other students in your class.

2. What is meant by the rate of breathing? What do you need to measure to find this rate? Before you start your experiment, try your method of measuring your breathing rate to make sure it works.

3. What sort of exercise will you do and for how long will you do it? If

you want to compare your results with those of others in your class, you will all need to agree to do exactly the same exercise.

4. Can you keep an eye on the time as you do the exercise or do you need someone to help you?
5. What time interval will you use to measure your breathing rate? For how long after the exercise ends will you carry on measuring your breathing rate?
6. How will you record your results? If you decide to use a table, you should draw it before you start your experiment and perhaps check with your teacher that it is suitable. Appendix A helps you design a table.

HEART AND CIRCULATION

19.1 Introduction

The heart and the blood circulation are the transport system of the body. Blood carries substances and heat around the body. Blood is forced around the body, through **arteries**, narrow **capillaries** and **veins**, by a muscular pump, the **heart**. Figure 19.1 shows that arteries always take blood from the heart to the capillaries of the lungs and to other parts of the body and that veins always take blood back to the heart. Blood flows from arteries to veins through capillaries.

19.2 Blood

Blood is a mixture of a pale yellow fluid called **plasma**, **red** and **white blood cells** and bits of cytoplasm called **blood platelets**. Figure 19.2 shows the composition of 100 cm^3 of blood separated into plasma and solids. Most of the substances carried to and from the different cells are dissolved in the plasma; lipids are also carried in the plasma but as very small droplets. Oxygen is carried in the red blood cells. Carbon dioxide is carried in both the plasma and red blood cells. The structure and functions of red and white blood cells are described in Section 14.1.

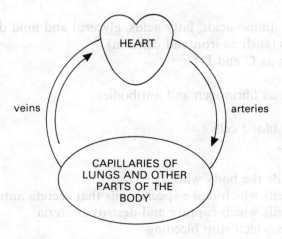

Figure 19.1 Transport in the human body

227

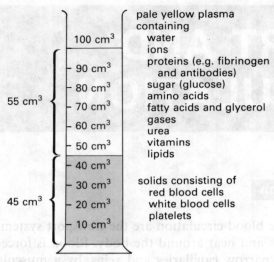

Figure 19.2   Contents of human blood

Blood platelets, small bits of cytoplasm, help to form blood clots at wounds and therefore to stop bleeding. When blood is exposed to air, the blood platelets release substances which help change the dissolved **fibrinogen** in the plasma into fine strands of **fibrin**. The blood cells are quickly trapped in a network of fibrin and form a **clot**.

Usually a clot is formed only when a blood vessel is damaged with the result that blood is exposed to air. But if the lining of a blood vessel is unhealthy, a clot can form on it. Such a clot, called a **thrombosis**, can partly or completely block a blood vessel. The harm this can cause is dealt with in Section 19.4.

Circulating in the blood transport system are

○ oxygen
○ carbon dioxide
○ water
○ food (glucose, amino acids, fatty acids, glycerol and lipid droplets)
○ inorganic ions (such as iron and calcium)
○ vitamins (such as C and D)
○ urea
○ proteins (such as fibrinogen and antibodies)
○ plasma
○ red and white blood cells
○ blood platelets
○ heat

The blood defends the body with

○ white blood cells which make special cells that secrete antibodies
○ white blood cells which capture and destroy bacteria
○ blood platelets which stop bleeding

228   Note that these white blood cells are of two different kinds.

Unit 18 explains that blood goes to the lungs after it has been to the other parts of the body. In the lungs the blood gives up carbon dioxide and gets a fresh supply of oxygen. After blood has been in the narrow capillaries of the body, it flows slowly and sluggishly in the veins until it returns to the heart, where it is pumped, with renewed pressure, to the lungs. After blood has been in the narrow capillaries of the lungs, it again flows slowly and sluggishly in the veins until it again returns to the heart, where it is pumped, with renewed pressure, to the other parts of the body.

The blood pressure to the lungs is lower than the blood pressure to the other parts of the body. Because the lungs are near to the heart, a low pressure is enough to pump blood to them. The air sacs of the lungs are fragile and thin-walled and would be damaged by high blood pressure. Much greater pressure is needed to force blood the longer distances to the head, the fingers and the toes.

Blood travels in a **double circulation**: it goes twice through the heart before it gets back to where it started. The heart is separated into left and right sides: blood from the right side is pumped at low pressure to the lungs; blood from the left side is pumped at high pressure to the other parts of the body. Blood from the lungs, which is now **oxygenated**, returns to the left side of the heart to be pumped to the other parts of the body; blood from the other parts of the body, which is now **deoxygenated**, returns to the right side of the heart to be pumped to the lungs. By the time blood gets back to the heart on either side, the pressure is almost zero.

This double circulation is shown in Figure 19.3. (The left side of the heart is shown on the right, and the right side is shown on the left: this is how you would see the heart of someone facing you.) Figure 19.3 also shows blood leaving the heart in arteries and returning to it in veins. Arteries are always thicker-walled and tougher than veins: they have to withstand a higher blood pressure. Capillaries have walls only one cell thick.

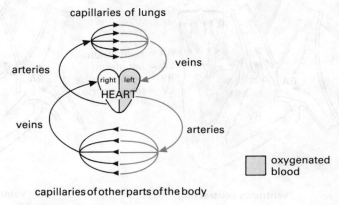

Figure 19.3   Double circulation

229

Not only is the heart divided into two completely separate left and right sides, but each side is itself divided into two parts: a thin-walled upper chamber, the **atrium**, which receives blood at near zero pressure from the veins, and a thick-walled lower chamber, the **ventricle**, which pumps blood out at higher pressure through the arteries. The heart therefore has four chambers: left and right atria (the plural of *atrium*) and left and right ventricles. Whereas the left and right sides are completely separate, each atrium leads by a narrow opening controlled by a valve into the ventricle on its own side.

Figure 19.4 shows from the inside how the human heart acts as a pump. The ventricles are thick-walled and muscular. When the muscles of a ventricle wall contract, the blood inside the ventricle is forced out into the arteries. It cannot go back into the atrium because the valve closes. When the muscles of the ventricle wall relax, the space inside the ventricle expands. Blood is then drawn into the ventricle from the atrium and, in turn, from the veins. Blood cannot come back from the artery because another valve that lies between the ventricle and the artery closes.

The two sides of the heart behave in exactly the same way at the same time. But there is an important difference between the two sides: the wall of the right ventricle is not as thick or as muscular as that of the left ventricle and so pumps the blood out at a much lower pressure.

As both ventricles relax, blood enters the heart from the veins and passes through the two atria and through the two valves. As both ventricles contract, blood is forced out of the right ventricle into the artery that leads to the lungs and out of the left ventricle into the major artery that leads to the other arteries and then to the other parts of the body. All this is shown in Figure 19.4.

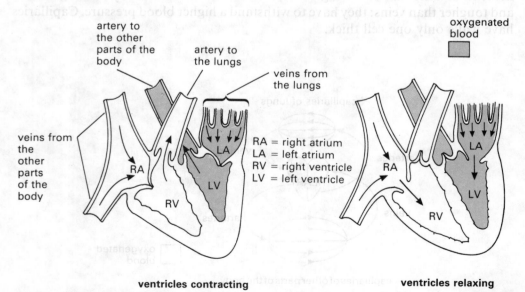

oxygenated blood

RA = right atrium
LA = left atrium
RV = right ventricle
LV = left ventricle

ventricles contracting   **ventricles relaxing**

230   Figure 19.4   Human heart and blood flow

Normally the heart valves stop all backward flow of blood, but some babies are born with faulty valves and the valves of older people sometimes become faulty due to hardening of the valve tissue. When the valves are faulty, they leak blood backwards and the heart cannot work so well. 'Open-heart surgery', during which the blood misses out the heart and lungs and goes instead to a machine which does their work, makes it possible for surgeons to cut open the heart to repair, and even to replace, faulty valves.

So far the diagrams have not shown you what the heart looks like from the outside or where it lies in the human body. Figure 19.5 is a cut-away diagram of the chest and abdominal cavities. It shows the heart lying slightly on its side above the **diaphragm** (the muscle that separates the chest from the abdomen) and between the two lungs. The diagram also shows some of the larger blood vessels and the two **kidneys**, which lie against the back wall of the abdomen.

(a) The pulse

The two stages in the heartbeat – contraction and relaxation of the muscles of

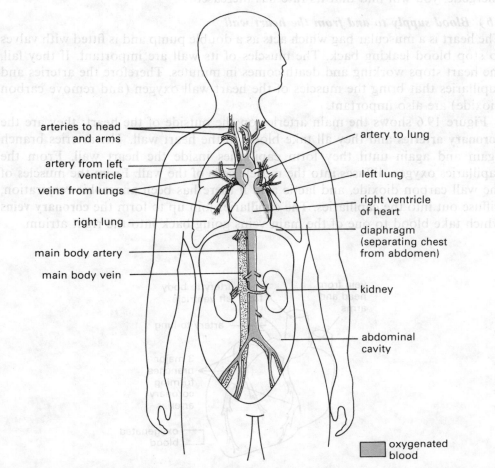

arteries to head
and arms

artery from left
ventricle

veins from lungs

right lung

main body artery

main body vein

artery to lung

left lung

right ventricle
of heart

diaphragm
(separating chest
from abdomen)

kidney

abdominal
cavity

oxygenated
blood

Figure 19.5   Major organs and blood vessels in the chest and abdomen (gut removed)   231

the ventricles – produce a **pulse** which can be felt in the arteries. In certain parts of the body (for example, at the wrist, in the neck and at the side of the forehead), you can feel the pulse in your own arteries. Sometimes you can see the pulse in the artery in the forehead. The pulse is produced by the pressure of blood as the left ventricle contracts causing the artery to expand. This is followed by a pause when the ventricle relaxes and the artery shrinks.

When you 'take your pulse', you are counting the number of contractions of the left ventricle. You are also indirectly counting the number of contractions of the right ventricle because the two always work together. The pulse rate is therefore the same as the heartbeat rate. Experiment 19.1 tells you how to measure the pulse rate.

As soon as you start to do anything energetic, more oxygen is needed in the muscles to release energy in respiration. More waste carbon dioxide is produced in the muscles and this must be removed. The circulating blood has to speed up so that more blood arrives at the muscles to bring the oxygen and take the carbon dioxide away. The ventricles therefore beat more often: the heartbeat rate speeds up. If you take your pulse when you are doing, or have just done, something energetic, you will find that its rate has increased.

(b) Blood supply to and from the heart wall

The heart is a muscular bag which acts as a double pump and is fitted with valves to stop blood leaking back. The muscles of its wall are important. If they fail, the heart stops working and death comes in minutes. Therefore the arteries and capillaries that bring the muscles of the heart wall oxygen (and remove carbon dioxide) are also important.

Figure 19.6 shows the main arteries on the outside of the heart: they are the **coronary arteries** and they all take blood to the heart wall. The arteries branch again and again until they form capillaries inside the heart wall. From the capillaries oxygen diffuses into the muscle cells of the wall. From the muscles of the wall carbon dioxide, and lactic acid if there has been anaerobic respiration, diffuse out into the capillaries. The capillaries join up to form the **coronary veins** which take blood to one of the main veins going back into the right atrium.

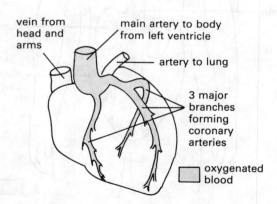

Figure 19.6   Coronary arteries

In a new-born baby the inside lining of the coronary arteries is perfectly smooth. Blood flows easily through these arteries. With age the lining becomes thickened and irregular, mainly because patches of lipid form on the inside of the wall. These cause a furring-up of the arteries (called **atherosclerosis**). Although it occurs in arteries all over the body, it matters a great deal in the coronary arteries because the heart muscles must never stop working. Reduced blood flow to the heart muscles, the result of furred-up coronary arteries, causes **coronary heart disease** (often called simply **CHD**).

Figure 19.7 shows how badly the arteries can get furred up. If such furring-up reduces the blood supply (and hence the oxygen supply) through the coronary arteries, the result may be a gripping pain in the chest called **angina**. Once an artery is furred up, a clot is more likely, and a clot may block an artery so completely that no blood passes. A blood clot in a coronary artery is called a **coronary thrombosis** (sometimes simply a 'coronary') or a **heart attack**. It causes an extreme gripping pain in the chest which can spread to the neck, jaw and arm.

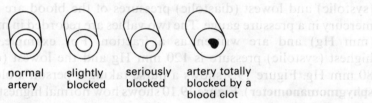

| normal artery | slightly blocked | seriously blocked | artery totally blocked by a blood clot |

Figure 19.7   Furred-up arteries

When there is a blood clot in a coronary artery, the muscle beyond that point can no longer contract and it dies. Figure 19.8 shows three examples of blocked coronary arteries. If the blockage occurs at **A**, a small artery, only a small region of heart muscle dies; if it occurs at **B**, much more heart muscle dies; if it occurs at **C**, so much heart muscle dies that the whole heart may stop beating with the result that the person dies. This is called a **cardiac arrest**. (*Cardiac* means *to do with the heart*.)

A study at the University of Nashville in the United States has shown that

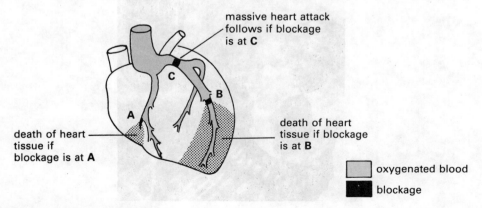

massive heart attack follows if blockage is at **C**

death of heart tissue if blockage is at **A**

death of heart tissue if blockage is at **B**

oxygenated blood

blockage

Figure 19.8   Heart attacks

233

regular cigarette smokers have abnormal blood platelets that form blood clots more readily than normal ones.

(c) Blood pressure

Blood pressure is the name given to the force exerted on the blood in the major arteries to keep it moving around the body. The greatest force, called the **systolic pressure**, occurs when the left ventricle contracts and pushes blood into the main artery (and thence to other arteries throughout the body). This is what you feel in the pulse. When the left ventricle relaxes, blood still flows in the arteries but the pressure, called the **diastolic pressure**, is much lower. The lower diastolic pressure is caused by the shrinking of the artery walls after they have been stretched by the systolic pressure. The artery walls are elastic or they would not stretch and shrink.

Blood pressure varies a great deal in different people, but it tends to get higher as people get older. Exactly why this happens is not clear, but furring-up of the arteries is one cause. To measure blood pressure a doctor or nurse wraps a black rubber bag around the patient's arm and blows it up with air. Both the highest (systolic) and lowest (diastolic) pressures of the blood are shown by heights of mercury in a pressure gauge. The two values are recorded in millimetres of mercury (mm Hg) and are written as a fraction. For example, 120/80 means the highest (systolic) pressure is 120 mm Hg and the lowest (diastolic) pressure is 80 mm Hg. Figure 19.9 shows a nurse taking a person's blood pressure (with a **sphygmomanometer**). Figure 19.10 shows how normal highest (systolic) and lowest (diastolic) blood pressures vary with age.

(d) Exercise and the heart

Section 17.2 describes how regular exercise can add years to a person's life: it keeps calcium in the bones and improves muscle efficiency. Section 18.4 describes

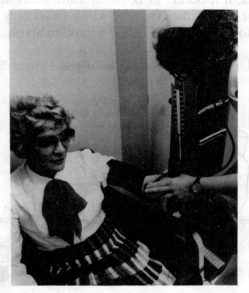

234   Figure 19.9   Taking a person's blood pressure

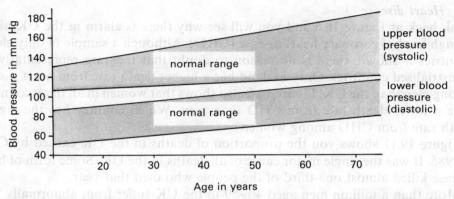

Figure 19.10 Normal highest (systolic) and lowest (diastolic) blood pressures

how regular exercise increases the volume of air a person can breathe in at one breath and the number of breaths a person can comfortably take per minute. Regular exercise is also helpful in keeping down the amount of stored fat. Exercise always uses up energy that has recently been eaten in food and may otherwise be converted to fat. In someone who is not eating too much food, exercise will use up some of the fat stores in the body.

Regular exercise can do a lot for the heart too. More muscles develop in the walls of the heart and the heart gets bigger. The ventricles also get bigger and empty better at each heartbeat: more blood is therefore pumped out every time the heart beats. The wall of the major artery from the left ventricle remains elastic and free of fatty deposits. Table 19.1 shows you what a difference regular exercise can make.

The data in Table 19.1 show that the work done by both the trained athlete and the untrained person was roughly the same (because they needed the same amount of oxygen). But the trained athlete's heart did not need to contract so often because it could pump out more blood at each heartbeat. Less blood was needed each minute (19.5 dm^3 compared with 21.6 dm^3) in the trained athlete because of a better blood supply to the muscles which did the work and a higher oxygen content in the blood.

There is also evidence that exercise reduces the amount of lipid lining the walls of blood vessels.

Table 19.1 Heart data

	Trained athlete	Untrained person
Oxygen needed for task in dm^3 per minute	3.023	3.023
Heartbeats per minute	125	180
Volume of blood pumped per heartbeat in dm^3	0.156	0.120
Volume of blood pumped per minute in dm^3	19.5	21.6

235

(e) Heart disease

Look back at Figure 16.6 and you will see why there is alarm in the UK about its high level of coronary heart disease (CHD). Although a sample of only eleven countries is shown, there is no reason to doubt that they are representative of industrialised countries. Only Finland has a higher death rate from heart disease among men than the UK. Figure 16.6 also shows that women in all these countries have a lower death rate from CHD than men. Which country has the highest death rate from CHD among women?

Figure 19.11 shows you the proportion of deaths in the UK caused by CHD in 1985. It was the single major cause of all deaths in the UK. Some form of heart disease killed almost one-third of the people who died that year.

More than a million men aged 40–59 in the UK suffer from abnormally high blood pressure (called **hypertension**). They are three times as likely to suffer serious heart disease as the twenty per cent of the population who have the lowest blood pressure.

A sample of 735 60-year-old men in the United States was divided into five equal groups of 147 according to the percentage of fat in their body mass. Figure 19.12 is a histogram showing the percentage of each group being treated for high blood pressure. Treatment for high blood pressure was more common in the men whose body mass was made up of a large proportion of fat. But some thin men also suffered from high blood pressure.

The following are known or believed to increase the risk of coronary heart disease:

o smoking cigarettes
o high blood pressure
o high concentrations of blood cholesterol
o overweight
o lack of exercise (not proved)

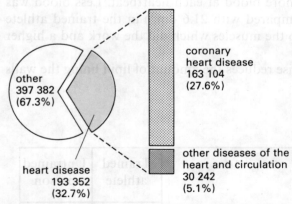

Figure 19.11   Proportion of deaths due to heart disease in 1985 (source: *OPCS Monitor*, 1986)

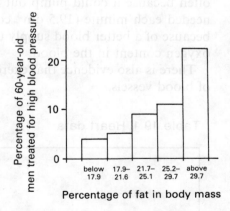

Figure 19.12   Relation between percentage of fat in body mass and treatment for high blood pressure (source: *Report of the Royal College of Physicians*, 1983)

236

So what should you do to reduce the risk of coronary heart disease?

○ Keep your body weight within normal limits

○ Take exercise at least three times a week

○ Give up smoking or, better still, don't start

○ Eat few saturated fats, if necessary replacing some saturated fats with unsaturated ones (see Section 16.4)

○ Eat plenty of vegetables and fruit

○ Eat little salt and few processed foods (which contain a lot of salt)

If you have a parent, brother or sister who has had heart disease, have a blood-cholesterol test and take the advice that the specialists give you.

Questions

Q 19.1. Make a table like the one shown below but larger. Fill in the details for red and white blood cells and for blood platelets (see Sections 14.1 and 19.2).

	Red blood cells	White blood cells	Blood platelets
Structure Functions			

Q 19.2. Make a table like the one shown below but larger. Fill in the details for arteries, capillaries and veins (see Section 19.3).

	Arteries	Capillaries	Veins
Direction of blood flow Thickness of wall Blood pressure			

Q 19.3. Look back at Section 16.4 and Figure 16.5.

(a) Explain how Americans changed their diet and reduced their risk of heart disease.

(b) What else did the Americans do to reduce their risk of heart disease?

Q 19.4. Look at Figure 19.13 showing the risk of coronary heart disease (CHD) in non-smokers (those who have never smoked), smokers and ex-smokers. Answer the questions that follow.

(a) Which group is at the greatest risk of death from CHD?

(b) Which group is at slightly less risk of death from CHD than those who have never smoked?

(c) There are other unexpected data in this graph. What are they?

237

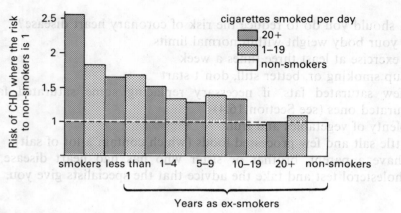

Figure 19.13   Graph for Question 19.4

Experiment 19.1 To test the hypothesis that exercise affects the pulse rate

Note to teachers

Many students are concerned about their own fitness and like to have a way of measuring it. The method below describes a standard test used in sports science to measure fitness. Some students may like to develop this experiment into a longer-term study of the effect of continued exercise on their recovery rate. You may like to encourage students who are less interested in physical activity to devise a simpler experiment of their own.

> **Do not let students do this Experiment if they have medical problems which make it unsafe for them to exercise.**

Tell students some time before they carry out this experiment that they need to bring appropriate footwear and clothing. You should perhaps warn colleagues holding classes nearby that this experiment may be noisy.

Materials needed by each pair of students
1 stopclock
1 sturdy bench about 45 cm high (for males)
1 sturdy bench about 40 cm high (for females)

Instructions to students
1. Choose a partner with whom to work. Each of you will help the other to keep to a rhythm while exercising.
2. Copy the table.

Time	Pulse in beats per 30 s
Before exercise begins 1 minute after exercise 2 minutes after exercise 3 minutes after exercise	

3. Turn the palm of one of your hands towards your face. Place the fingers of your other hand on the wrist facing you about a centimetre below your hand on the same side as your thumb. You should be able to feel your pulse. If not, try again or ask for help. Use the stopclock to count your pulse during a 30-second period and record the value in the first space in your table.

4. Stand in front of the bench ready to step on and off it. Your partner will tell you when to begin and will start the stopclock at the same time.

5. Step on and off the bench at a steady rate of 30 steps per minute for a total of 4 minutes (120 steps). Your partner will help by counting your steps aloud, using the stopclock to ensure accurate timing. If you fall behind or go ahead of the rate of 30 steps per minute, stop, take a short rest and start all over again.

6. At the end of the 4 minutes sit on the bench. Count your pulse during three 30-second periods, starting your counts exactly 1 minute, 2 minutes and 3 minutes after you finished your exercise. Your partner will tell you when it is time to begin each of the 30-second counts and will record the results in your table.

7. Now swap roles with your partner and repeat the experiment.

Interpretation of results

1. Compare your pulse rate before and immediately after the exercise. What effect did exercise have on your pulse rate? Explain why your pulse rate changed.

2. Did your pulse rate change during the 3 minutes after exercise? If so, can you explain why?

3. Your data can be used to calculate a measure of your fitness called the CR score. CR stands for cardiorespiratory: *cardio* refers to your heartbeat or pulse rate and *respiratory* to your breathing. To calculate your CR score you must substitute your own values in the equation:

$$\text{CR score} = \frac{(\text{total time of exercise in seconds}) \times 100}{2 \times (\text{total of your three pulse counts})}$$

Because you exercised for 4 minutes, the equation becomes:

$$\text{CR score} = \frac{240 \times 100}{2 \times (\text{total of your three pulse counts})}$$

$$= \frac{12\,000}{(\text{total of your three pulse counts})}$$

239

Your CR score shows your fitness:

90 +   superior
80–89 excellent
70–79 good
60–69 fair
50–59 poor
under 50 very poor

TEMPERATURE CONTROL AND HOMEOSTASIS

The human body works best at an internal temperature at about 37°C. In most parts of the world the outside temperature is rarely as high as this. We think it is hot when the surrounding temperature is higher than 30°C. At 30°C a naked inactive person is comfortable because the heat lost from the body is matched by the heat released in respiration of the cells.

During a 24-hour day 95 per cent of the energy released by the cells in respiration is in the form of heat. Because humans wear clothes, this is often more heat than the body needs to keep its temperature at about 37°C despite a continuous loss of heat from the skin and in the gases we breathe out and a further loss of heat in urine and faeces. The bodies of people in good health keep their temperature at about 37°C by a finely tuned control system.

This Unit describes natural ways in which our bodies cut down heat losses when the temperature about us is cold. If we are still cold, we put on more clothes or we put on fires or other artificial heating. Unless the outside temperature is very hot, our bodies do not, as we often imagine, gain heat from clothes and fires. All that clothes and fires (and the sun) usually do is reduce the rate at which we lose the heat we have produced for ourselves by respiration.

20.2 Temperature control

Keeping the internal body temperature at about 37°C involves not only producing heat when it is needed, but transporting it throughout the body, detecting when the temperature is too high or too low and setting the corrective processes in motion.

(a) Heat production

Respiration is the heat-producing process in the body. Most heat is produced in the muscles, in the liver and in certain special cells, called **brown-fat cells**, under the skin at the back of the neck and between the shoulder blades. The table in Section 16.4 shows that the body releases energy even when we are asleep. As soon as we take even gentle exercise, more respiration takes place and more heat is released. If we are very cold, the increase in respiration in the muscles to give heat is so great that we feel the resulting muscle contractions as **shivering**.

(b) Heat transport

Blood carries heat away from the places where respiration is vigorous. As it passes through colder regions of the body, it gives up some of its heat to them.

(c) Corrective processes

A part of the **brain** (the **hypothalamus**) can detect whether the blood flowing through it is too hot or too cold. It also receives **impulses**, via **nerves** from **sense cells** in the skin, which give information about the skin's temperature. Figure 20.1 shows diagrammatically how the brain controls temperature both when the body is too hot and when it is too cold.

If the temperature is too high, the brain sets in motion, through impulses via nerves and through reduced secretion of **hormones**, the corrective processes:
○ increased heat loss from the skin
○ reduced heat production by respiration in the muscles and brown-fat cells
If the temperature is too low, the brain sets in motion, through impulses via nerves and through increased secretion of hormones, the corrective processes:
○ reduced heat loss from the skin
○ increased heat production by respiration in the muscles and brown-fat cells

(d) Heat loss

The body loses heat by
○ radiation from the skin
○ convection when heated air is moved away from the skin
○ conduction when objects that touch the skin are heated by it
○ evaporation from the skin, the lungs and the air passages from the lungs

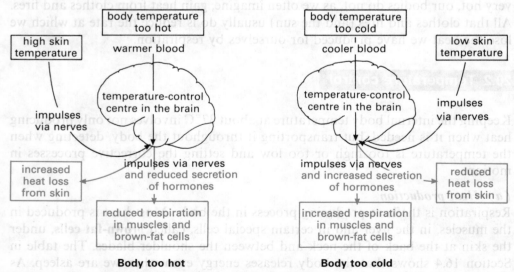

242   Figure 20.1   Temperature control

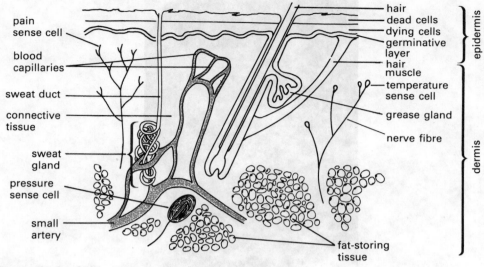

Figure 20.2   Skin section

There is evaporation all the time from the water in the skin cells and from the wet lung surfaces and the wet air passages leading from the lungs. The greatest heat loss occurs when the water in **sweat** (from the **sweat glands**) is evaporated from the surface. Energy needed to change water to vapour (called **specific latent heat of vaporisation**) is obtained from heat in the skin. The skin is cooled, blood in the skin is cooled and cooled blood is carried around the body. The function of sweating is to cool us when we are hot.

Figure 20.2 shows a section through the skin, including those parts that are important in controlling temperature losses in humans. The **temperature sense cells** send impulses to the brain giving information about the temperature of the skin. Fat-storing tissue under the skin provides **insulation**: it reduces heat loss but cannot prevent it.

(i) Increasing heat loss

If the body is too hot, the **sweat glands** extract water, urea and inorganic ions from the blood capillaries that surround them to form **sweat**. Sweat pours along the duct and out on to the skin surface, where its evaporation cools the skin and the blood in the skin. Figure 20.3 shows sweat on the face of an athlete after vigorous exercise. So much sweat formed that it could not evaporate quickly enough. Figure 20.4 shows how sweat glands release sweat on to the skin surface.

At the same time the muscles in the walls of the small arteries taking blood to the skin relax, the arteries get wider (**dilate**) and more blood flows through them. (The process is called **vasodilation**.) As a result, more blood flows in the capillaries of the skin and more blood gets cooled. This is shown in Figure 20.4. When people are too hot, their skin is darker because there is more blood in it.   243

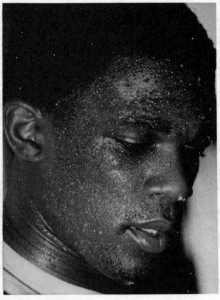

Figure 20.3   Sweating after vigorous exercise

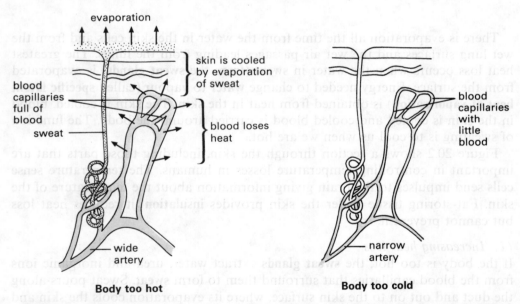

Figure 20.4   Widening and narrowing of skin arteries (vasodilation and vasoconstriction)

(ii) Reducing heat loss

If the body is too cold, no sweat forms. The muscles in the walls of the small arteries taking blood to the skin contract, the arteries get narrower (**constrict**) and less blood flows through them. (The process is called **vasoconstriction**). As a result, less blood flows in the capillaries of the skin and less blood gets cooled. This is shown in Figure 20.4. When people are too cold, their skin is paler because there is less blood in it.

(iii) Uncontrollable heat loss

Because the surrounding temperature is usually lower than that of the body, there is usually some heat loss at the skin surface by radiation, convection, conduction and evaporation. Because the surface of the skin is not waterproof, there is always some water evaporating from the skin cells (even when there is no sweating). Figure 20.2 shows dead and dying cells at the skin surface: water evaporates from them too.

Humans have fine short hair all over their bodies. Some of their hair (particularly their head hair, pubic hair and under-arm hair) is longer and coarser. Most mammals are covered in long coarse hair which helps control heat losses by increasing and decreasing the depth of the hair layer. In humans hair plays no part in temperature control except that hair on the head traps air which reduces heat loss or prevents overheating in bright sunlight.

20.3 Hypothermia

When the body temperature, which should be about 37°C, falls below about 35°C, the temperature-control system breaks down. The body temperature cannot be quickly raised to 37°C without outside help. This condition is called **hypothermia** (the word means *under heat*).

Between a body temperature of 35°C and 25°C, hypothermia can usually be overcome if correct treatment is given. Below 25°C the treatment usually fails and the person dies.

Those most likely to suffer from hypothermia are old people and babies. Old people get it because they take little exercise and because in winter they may not be able to afford to heat their rooms well enough. Babies get it because their temperature-control system is not yet perfected and because their body surface (where heat is lost) is large in relation to their small volume.

But it is not only old people and babies who get hypothermia. Anyone can get it who is poorly protected in cold windy conditions, particularly if exhausted and wearing wet clothes. Anyone can get it after a long time in cold water. Hypothermia occurs when the body is unable to produce enough heat quickly enough to make up for the loss of heat from the skin surface.

People suffering from hypothermia feel miserably cold, their skin is pale (because the arteries to their skin are narrowed), their skin feels unnaturally cold to touch, they may shiver uncontrollably, their pulse and breathing rate slow down and may be difficult to detect at all, and their mental activity is dulled with the result that they may even become unconscious.

Treatment tries to prevent any more heat loss and gradually to increase the body temperature. Hypothermia is a serious condition which needs specialised treatment in hospital. Until this can be arranged the person must be

○ kept lying down
○ covered (except for the face) with anything (such as a blanket) which will prevent further heat loss
○ put under shelter if outdoors or kept in a warm room if indoors
○ given a warm sweet drink (if conscious)

If medical help cannot be got quickly, the person must be warmed with hot-water bottles next to the trunk but not next to the limbs.

Do not

o put hot-water bottles next to the person's limbs
o give the person alcohol
o rub or massage the limbs
o encourage the person to stand or take exercise

These last four measures all increase heat loss by taking blood to the limbs and skin, which are the coldest parts of the body. It is important to heat the trunk first. Once the organs in the trunk have warmed up, they will release heat which the blood will take to the limbs.

20.4 Homeostasis

Homeostasis is the process of keeping conditions in the body constant or within narrow limits. Temperature control is one example of homeostasis. There are many others: for example, the concentration of most of the substances in blood plasma listed in Figure 19.2 must be kept within safe limits or they would harm the body. In Section 18.3 the removal of excess carbon dioxide from the blood in the lungs, by stronger and quicker breathing movements of the chest muscles, is another example of homeostasis. In Section 20.5 the removal of excess water and urea from the blood in the kidney is yet another example.

Homeostasis works like non-biological control systems. There is **feedback**: the results of an adjusting action are fed back and detected by the system to allow further necessary adjustments to be made. If the adjustments reduce the difference between what is detected and what is needed, the feedback is called **negative feedback**. Thus negative feedback in spite of its name is a good thing.

Suppose the body-core temperature falls to 36.0°C. The control centre in the brain detects this and sets corrective processes in motion (such as increased muscle activity, reduced sweating and narrowing of the arteries taking blood to the skin). Because all these processes take time to work, the temperature overshoots 37.0°C to, say, 37.5°C. The brain detects this and sets corrective processes in motion (such as reduced muscle activity, increased sweating and widening of the arteries to the skin). The overshooting will be less this time, perhaps to 36.75°C. Provided that the brain moderates the corrective process each time, so that the overshooting gradually decreases, the body temperature will end up at 37.0°C. (In control systems, making these attempts to find the correct result is called **hunting**.)

With constant changes in body activity and in surrounding conditions, the temperature-control system is continually making adjustments. The same is true of all the control systems in the body. Homeostasis is a never-ending process in a living person.

20.5 Removal of waste products

246   The important waste products produced in the body are

○ carbon dioxide and water from respiration
○ urea from the breakdown of excess amino acids

Carbon dioxide and urea are **toxic** (poisonous) in high concentrations. Too much water is dangerous because it dilutes first the blood and then the tissues.

(a) Carbon dioxide

The control system that regulates the concentration of carbon dioxide in the blood is hinted at in Section 18.3. Its concentration is detected by blood vessels in the neck but, as in the temperature-regulation system, the centre from which the carbon-dioxide concentration is controlled is in the brain. From here impulses are passed via nerves to set in motion the corrective processes: if the carbon-dioxide concentration is too high, breathing becomes deeper and faster; if the carbon-dioxide concentration is too low, breathing becomes shallower and slower.

(b) Urea

The body cannot store amino acids or protein. Excess protein that is eaten, or protein broken down in the body as a source of energy, is changed first into amino acids and then into a toxic nitrogen compound, **urea**, and a harmless non-nitrogen compound which is used as a source of energy. The process is called **deamination** and takes place in the liver. Figure 20.5 sums up this process of deamination.

Urea passes into the blood and is extracted from blood in the kidneys.

(c) Water

Water from respiration and water absorbed from food and drink in the gut both pass into the blood. Water gains dilute the blood. Water is lost from the body in: evaporation from the skin cells, from the moist lung surfaces and from the moist air passages leading from the lungs; sweat (which also evaporates); tears and other secretions. Water loss concentrates the blood.

20.6 The kidneys

Figure 19.5 shows the two **kidneys** lying against the back wall in the abdomen. It also shows two major blood vessels sending branches to each kidney. Each

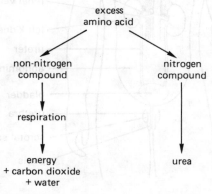

Figure 20.5   Deamination

branch from the main artery is a **renal artery**; each branch to the main vein is a **renal vein**. (*Renal* means *to do with the kidneys*.)

The brain detects the concentrations of the substances dissolved in the water in the blood and sets in motion the corrective processes in the kidneys. The kidneys remove from the blood both water and substances dissolved in it in the exact quantities necessary to keep each of the dissolved substances at the correct concentration (another example of homeostasis).

Figure 20.6 shows further details of the **urinary system**, which removes excess urea and water from the body. The **ureter** carries **urine**, which is formed from excess urea and water in the kidney, down to the **bladder**. In the bladder urine is stored to be passed out from the body through the **urethra** every few hours. In a female the urethra, the passage from the bladder to the outside, is shorter.

To make urine, the kidney works as a filtering system. Blood goes into the kidney by a renal artery which then divides into over a million capillaries in each kidney. About ten per cent of the blood going into the kidney, which is about a fifth of the fluid plasma, is filtered out. The filtering is done by over a million tiny filters each of which has blood brought to it by a small capillary. They are such fine filters that only water and substances in true solution (called the **filtrate**) pass through. Blood cells, platelets and large protein molecules all stay in the capillaries. Not all the water and dissolved substances in the plasma are filtered out: four-fifths stays in the blood capillaries.

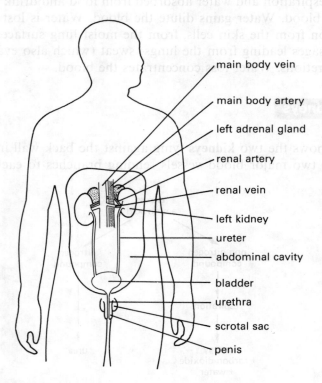

248   Figure 20.6   Urinary system

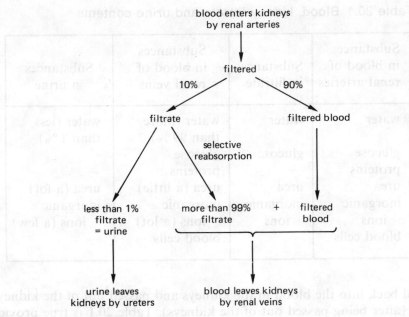

Figure 20.7 The work of the kidneys

Before the blood capillaries join up again to form the renal veins, which take blood from the kidneys, they pass over and between fine single-cell-walled tubes that lead from the filters. This is where the corrective processes of homeostasis are carried out. Nearly everything (more than 99 per cent) that is filtered out of the blood (including water) is absorbed back into the blood: but, if there is enough or too much of any substance in the blood, it is not absorbed back. The process is called **selective reabsorption**. The excess, now called urine, passes out of the fine tubes in the kidney and drains down the ureter to the bladder. This process is summed-up in Figure 20.7.

In 1 dm^3 of urine there are about

> 960 g water
> 20 g urea
> 12 g inorganic ions
> traces of other substances

The exact amount of each substance varies with the amount of water drunk, the amount of protein eaten and the amounts of different inorganic ions eaten. Which other substances there are and how much of them there are also depend on what has been eaten, because some substances, for example certain drugs which cannot be broken down in the body, are got rid of in urine. Urine sometimes contains traces of hormones, for example reproductive hormones which are produced in large amounts when a woman becomes pregnant. A motorist's blood-alcohol level can be determined from a urine sample.

Table 20.1 shows some of the important substances in the blood of the renal arteries, in the filtrate in the kidneys, in the blood of the renal veins (after being 249

Table 20.1 Blood, kidney filtrate and urine contents

Substances in blood of renal arteries	Substances in filtrate	Substances in blood of renal veins	Substances in urine
water	water	water (more than 99%)	water (less than 1%)
glucose	glucose	glucose	—
proteins	—	proteins	—
urea	urea	urea (a little)	urea (a lot)
inorganic ions	inorganic ions	inorganic ions (a lot)	inorganic ions (a few)
blood cells	—	blood cells	—

absorbed back into the blood in the kidneys and passed out of the kidneys) and in urine (after being passed out of the kidneys). Table 20.1 is true provided the kidneys are healthy. Substances showing up in urine may be symptoms of illness: for example, glucose in urine is a symptom of **diabetes**.

20.7 Toxic substances taken into the body

Waste products are not the only toxins (poisons) that the body has to deal with: some toxins are taken into the body in food and drink. Some toxins the body can deal with; others it cannot. Those it cannot deal with may be got rid of in vomit, urine or faeces; if they cannot be got rid of, they can kill us. *Staphylococci* bacteria, a common cause of food poisoning, produce a toxin which is not destroyed by cooking and which makes us feel ill but does not usually kill us.

Alcohol is a good example of a toxin which can make us feel ill, which can kill us if a lot is drunk quickly, but which can be broken down in the body if it is drunk slowly. In Section 5.4 you read that ethanol (alcohol) is a toxic end-product of anaerobic respiration of sugar in the cells of yeast. Alcohol contains only carbon, hydrogen and oxygen, and it still contains most of the energy in the original sugar. Provided that it is not drunk too quickly, it is carried in blood to the liver, where it is broken down in aerobic respiration to release energy, carbon dioxide and water. If alcohol is drunk faster than the rate at which the liver can break it down, liver cells are poisoned by the toxin and are replaced by scar tissue. Eventually this results in **cirrhosis** of the liver, from which heavy drinkers may die.

Questions

Q 20.1. Explain how and where
(a) heat is produced in the body
(b) heat is lost from the body.

250

Q 20.2. Explain how

(a) we can lose heat in an air temperature of 40°C

(b) a fan can cool us in an air temperature of 40°C.

Q 20.3. Explain why

(a) we find hot damp air more uncomfortable than hot dry air at the same temperature

(b) we should drink plenty of (clean) water in a hot dry climate.

Q 20.4. About twenty people a year die as a result of unexpected bad weather on the mountains and moors of the UK. Suggest what may cause these deaths.

Q 20.5. Table 20.2 gives information about a resting adult. Answer the following questions. (A calculator will help you.)

Table 20.2 Resting adult

Total volume of blood in the body	5.00 dm^3
Volume of blood leaving the left ventricle at each heartbeat	0.07 dm^3
Number of heartbeats per minute	70
Volume of blood passing into the kidneys per minute	1.20 dm^3
Volume of filtrate extracted in the kidneys per minute	0.12 dm^3
Volume of urine leaving the kidneys per day (24 hours)	1.50 dm^3

(a) (i) How much blood went through the left ventricle per minute?

   (ii) Compare your answer to (i) with the total amount of blood in the body. By how much is the total amount of blood in the body greater or smaller?

(b) What proportion of the total volume of blood in the body passed through the kidneys in one minute?

(c) What proportion of blood passing into the kidney was filtered out?

(d) (i) How much blood was filtered out in a day?

   (ii) How much of the filtrate (what is filtered out) was reabsorbed in a day?

NERVOUS AND HORMONAL CONTROL; DRUGS

21.1 Introduction

In order to keep healthy and safe the body must detect changes both inside and outside itself and make the correct **responses**. Section 20.2 explains in detail how the body detects and responds to changes in temperature, both inside and outside itself. Sections 18.2, 20.5 and 20.6 deal with the detection and removal of excess carbon dioxide and water in the blood. The ability to detect changes and respond to them in different parts of the body involves **co-ordination**, making sure that everything is working together in the right way. Co-ordination is made possible by **nervous impulses** via **nerves** or by **hormones** that travel in the blood or by both together.

21.2 Nervous impulses: the pupil reflex

The black centre of the coloured part of the eye is the **pupil**. The coloured ring around it is the **iris**. The pupil is really a hole in the iris through which light gets into the eye to form the images that we see. Bright light can damage the inside of the eye. The pupil diameter is continually adjusted, by muscles in the iris, so that only the right amount of light enters the eye. You can see for yourself in a mirror that the pupil is wider in dim light and narrower in bright light. Figure 21.1 shows how this happens.

The two sets of muscles in the iris are **antagonistic**. When one set **contracts**, it stretches the other set, which is **relaxed**. Antagonistic muscles are described in Section 17.1. The contraction and relaxation of the two sets of muscles are controlled by nervous impulses (electrical signals) that pass along nerve fibres from cells in the brain. (Bundles of nerve fibres form nerves.) We have no conscious control over the iris: its adjustments are **automatic**. Bright light closes the iris and dim light opens it. These are **reflex actions**.

When bright light enters the eye, impulses pass the information about its brightness via nerves to the brain. Automatically, without our thinking about it or being able to stop it, impulses pass from the brain along nerves to the two sets of iris muscles. The iris closes to allow just the right amount of light into the eye. You can probably guess what happens when the light gets dimmer: impulses

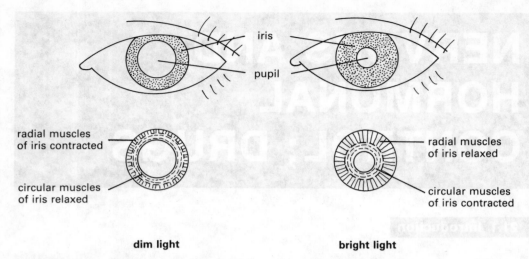

iris

pupil

radial muscles
of iris contracted

radial muscles
of iris relaxed

circular muscles
of iris relaxed

circular muscles
of iris contracted

dim light

bright light

Figure 21.1 Increasing and decreasing the pupil diameter

pass the information about its dimness via nerves to the brain; impulses from the brain along nerves to the two sets of iris muscles bring about the increase in the diameter of the pupil to allow just the right amount of light into the eye.

The pupil reflex is an example of nervous co-ordination. These nerves to and from the brain form a **reflex pathway** for the nerve impulses. It is called a **reflex** because the **stimulus** or change always results in a similar automatic response, one not under conscious control.

21.3 Hormones

Hormones are chemicals produced in response to a stimulus in one part of the body which have effects in other parts, called the **target organs**. Hormones are secreted from special glands straight into the blood, not into ducts (tubes). These glands are called **ductless**. Other glands, such as the sweat glands and grease glands found in the skin, do have ducts. Blood carries the hormones to the target organs. Dozens of different hormones are known to act in the human body. One you may have heard of is **adrenaline**. Adrenaline is secreted by the two adrenal glands. In Figure 20.6, you can see that there is one adrenal gland above each kidney. Small amounts of adrenaline are always present in the blood. Adrenaline is known as the fight-or-flight hormone because it increases the efficiency of fighting or escaping. In times of danger, nerve impulses from the brain stimulate the adrenals to pour out large amounts of adrenaline. Anxiety and excitement also stimulate the secretion of extra adrenaline.

Adrenaline

○ increases the glucose concentration in the blood by converting glycogen in the liver to glucose

○ decreases the diameter of arteries in the skin and gut (vasoconstriction) (the person goes pale)

254

- increases the diameter of arteries in the muscles and brain (vasodilation) (with the result that these arteries can receive diverted blood from the skin and gut)
- increases the rate of the heartbeat
- increases the blood pressure
- increases the breathing rate
- increases the pupil diameter

Look at this list. Can you see that each item increases the efficiency of fighting or escaping? The blood gets more glucose. Some blood is withdrawn from the skin and the gut and more is sent to the muscles and brain. An increased breathing rate gives blood passing over the lungs more oxygen, with the result that the blood going to the muscles and brain has more oxygen as well as more glucose, giving the person more energy and more alertness. Increases in the rate of heartbeat and blood pressure cause blood to be carried more quickly to muscles and brain. The increase in the pupil diameter lets more light into the eye so that the environment, and in particular movements in it, can be seen more clearly. There are many target organs. The result is that the person can take in what is happening quickly, can decide quickly what to do and has the energy needed either to fight or run away.

21.4 Hormones and nervous impulses compared

The two co-ordination systems are well suited to their different tasks. They are compared in Table 21.1.

Adrenaline acts more quickly than, and its effects are not as long-lasting as those of, most other hormones.

21.5 Drugs

A **drug** is a 'substance taken to help recovery from sickness, to relieve symptoms or to change natural processes in the body'. **Drug misuse** or **drug abuse** is the use of a drug for a purpose that is not medically or socially acceptable.

Table 21.1

Hormone co-ordination system, e.g. adrenaline	Nervous co-ordination system, e.g. the pupil reflex
Involves ductless glands	Involves the brain or central nervous system
Uses chemicals	Uses electrical impulses
Passage via blood vessels	Passage via nerves
Effects are usually slow	Effects are quick
Effects are widespread	Effects are localised
Effects are usually long-lasting	Effects are short-lived

255

(a) Medicines

Drugs thought to be medically useful are called **medicines**. They can be bought in shops, though some have to be prescribed by a doctor. All medicines are drugs, but not all drugs are medicines.

(i) Aspirin

The occasional use of an **aspirin** to dull the pain of a headache or toothache is medically acceptable and is not misuse of a drug. Aspirin is a **pain-killer** or, more accurately, a **pain-reliever**. But no drug is harmless. In many people aspirin causes slight bleeding of the stomach lining; if many aspirins are taken, or if aspirins are taken together with alcohol, bleeding can be severe. Some people are **allergic** to aspirin, which means that, if they take it more than once, they may get a skin rash or an asthma attack or other unusual reactions.

(ii) Penicillin

Penicillin belongs to a group of medicines called **antibiotics**. We use antibiotics, drugs made by certain microorganisms, to kill other microorganisms. Section 7.6 deals with the production of penicillin from the fungus *Penicillium*. In 1941 it became the first antibiotic ever to be produced on a large scale. Now there are many other antibiotics. Penicillin kills growing bacteria by damaging their developing cell walls; it does not damage non-growing bacteria or resistant spores. Because penicillin acts by damaging cell walls, it is usually harmless to humans, because their cells do not have walls. It is effective against bacterial infections of the lungs and breathing passages, of the bladder and urinary passages, and of the eye and ear.

Some people are allergic to penicillin: it can produce skin rashes, swelling of the face and throat, fever and swollen joints. It can even cause death through a rapid fall in blood pressure and acute asthma. People do not have allergic reactions to penicillin the first time they take it. The more often they take it, the more likely they are to have allergic reactions and the more likely are the reactions to be severe.

(b) Non-medical use of drugs

Addictive drugs are those which cannot easily be given up by the people who use them. Drug **dependence** is the term for the desperate need felt by some drug users to go on taking a drug. If stopping produces unpleasant physical effects, such as intense shivering, aching muscles and bones, vomiting and diarrhoea, a **physical dependence** has developed. If stopping produces anxiety or depression, a **psychological dependence** has developed. Drug dependence often involves both. The term **addiction**, which has the same meaning as drug dependence, is no longer used by medical experts.

Drug misuse increased in the UK in the early 1980s. Two drugs in particular, **heroin** and **cocaine**, became easier to get. Both drugs are controlled in Class A under the Misuse of Drugs Act, which means it is illegal to posses them or supply them to someone else without a prescription. In 1985 the Home Office gave the number of registered addicts (mainly heroin users) as 8819 but believed that the real figure of heroin users was closer to 45 000. Figure 21.2 shows that the total of registered drug addicts rose slowly in the 1970s, quickly in the early 80s and has since fallen slightly.

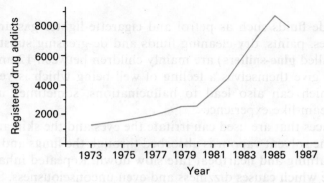

Figure 21.2   Registered drug addicts in the UK (source: Home Office, 1988)

(i) Heroin

Heroin is a white powder, produced from **morphine**, a compound which comes from the opium poppy. Heroin users sniff it, inhale it by smoking it or inject it. The user gets a feeling of relaxation and well-being because it removes stress and discomfort. Regular heroin users develop **tolerance**, which means they get used to it and therefore need more and more of it to get the same effect. Both physical and psychological dependence can be intense: withdrawal symptoms are severe; people who have become dependent on it say they need it just to stay 'normal'. Heroin is a powerful pain-killer with a medical use but is rarely prescribed by doctors today because it may lead to dependence and because other powerful pain-killers are available.

When addicts buy heroin illegally, they are often cheated. What they get is not pure heroin but heroin mixed with another substance such as glucose, flour, caffeine, talcum powder, chalk or brick dust. The heroin has been **diluted**. Some of the substances used to dilute heroin can poison heroin addicts who inject them and damage their blood vessels. Even when the heroin is pure, addicts often inject it under insanitary conditions with unsterile needles: they can then get **septicaemia** (blood poisoning) and abscesses. Shared needles can cause serious and fatal infections such as **AIDS** (see Section 22.7) and **hepatitis** (a virus infection of the liver).

(ii) Cocaine

Cocaine is a white powder that comes from the coca plant. It is a powerful stimulant which is sniffed or 'snorted' up the nose or is injected, sometimes mixed with heroin. It produces feelings of great physical strength and mental capability, of excitement and well-being and of indifference to pain or tiredness. Because it is quickly broken down in the body, doses have to be repeated if the feelings are to be kept. Repeated doses may lead to anxiety and panic, which in an extrene state can cause irregular breathing, nausea (feeling sick), confusion and even death from breathing failure or heart failure.

Intense psychological dependence can develop. After using cocaine, a person is tired, sleepy and depressed yet will want to use the drug again. There is not the physical dependence that there is with heroin.

257

(iii) Solvents

Solvents include fluids such as petrol and cigarette-lighter gas and compounds present in glues, paints, dry-cleaning fluids and de-greasing substances. Solvent users (often called **glue-sniffers**) are mainly children between 12 and 16 who sniff the solvent to give themselves a feeling of well-being which is very like getting drunk but which can also lead to hallucinations, sometimes in the form of frightening dream-like experiences.

The substances that are used can irritate the eyes and the skin around the nose and mouth. The solvent vapour is absorbed through the lungs and quickly affects the brain: breathing and heartbeat rate slow down. Repeated inhaling can result in an overdose which causes dizziness and even unconsciousness. Solvent-sniffers can die from falling unconscious in a dangerous place or from suffocating either by sniffing from an airtight plastic bag or by choking on their own vomit. There is no physical dependence. Long-term solvent abuse can lead to kidney and liver damage and to brain damage which particularly affects movement control.

(iv) Alcohol

Alcohol is a drug that can be sold legally in the UK, though only under licence and only to people aged 18 and over. It has killed far more people than heroin, cocaine and solvents. It is the third greatest killer (after heart disease and cancer) in the UK. Like the other drugs mentioned, it produces a feeling of well-being if it is taken in small amounts but is a dangerous substance which can even kill if taken in large amounts. Regular heavy alcohol drinkers suffer ulcers, circulation problems and brain damage as well as the better known liver damage (see Section 20.7).

Although alcohol is a sedative, it is often thought of as a stimulant. This is because its calming effect reduces anxiety and social shyness, allowing people to 'let themselves go' at a party or with friends. As the dose is increased, people become clumsy and slur their speech. Yet more alcohol results in staggering, double vision, loss of balance and finally unconsciousness. When alcohol is drunk, it is quickly absorbed into the blood and has an effect on the brain within five or ten minutes. The effect of a drink depends on

○ its strength ($\frac{1}{2}$ a pint of beer has roughly the same strength as a measure of spirits or a glass of wine)
○ the speed at which it is drunk
○ the amount of food in the stomach
○ the circumstances (a party atmosphere can increase the effect of drink)
○ the drinker's weight, personality and tolerance to alchol (a regular drinker has a greater tolerance than an occasional drinker)

Figure 21.3 shows the slowing of a man's responses after drinking three measures of whisky and $1\frac{1}{2}$ pints of beer. The time it took him to respond was measured by getting him to press a button as a light was switched on. He was tested twelve times before drinking and twelve times after drinking.

Driving with more than 80 mg of alcohol in 100 cm^3 of blood is illegal in the UK. A breath test gives a rough indication of the amount of alcohol in the blood: a blood test or urine test gives an accurate measurement.

The police in Michigan in the United States measured the blood-alcohol levels

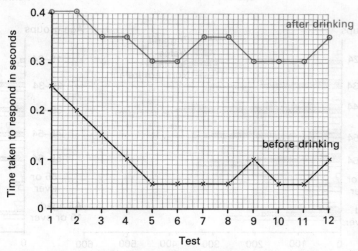

Figure 21.3 Time taken to respond before and after drinking alcohol

of every driver who had an accident. By stopping other drivers at the same time of day and at the same spots as the accidents took place, and by measuring their blood alcohol levels too, they were able to compare drivers who had had an accident with drivers who had not. Their results showed that at a level of 80 mg per 100 cm^3 drivers were twice as likely, and at a level of 150 mg per 100 cm^3 were ten times as likely, to have an accident as those who had no alcohol in their blood.

Over 90 per cent of the adult population drink alcohol at some time. On average men have the equivalent of $1\frac{1}{2}$ pints of beer a day and women the equivalent of $\frac{1}{2}$ pint. Fewer than one in twenty people regularly drink more than is thought to be medically safe (4 pints a day for men and 3 pints for women). Young adults drink more: about $1\frac{1}{2}$ times the average. In 1987 the Royal College of Psychiatrists pointed out that under-age drinking was commonplace and that the law against it was not being enforced: 60 per cent of teenagers under 18 had bought alcohol in a pub or off-licence.

Physical and psychological alcohol dependence can be severe. Sudden withdrawal from heavy drinking produces sweating, anxiety and trembling and can cause delirium and convulsions.

The social problems of heavy drinking (**alcoholism**) affect drinkers and their families. Alcoholics' lives are dominated by getting drinks: their other interests, families and work take second place. They may become ill from drinking and from neglecting their meals; they may have money problems because of the amount they spend on drinking; they may lose their jobs; their families may break up.

Heavy drinking in the UK is said to cost £2 billion a year in hospital bills, absence from work, road accidents and increased crime. Figure 21.4 shows the number of admissions to hospitals due to alcohol-related mental illnesses and alcohol misuse (including dependence). Data are given for men and women in England, Wales, Scotland and Northern Ireland per 100 000 in the population in each age group for the year 1981. Question 21.3 will help you compare the size   259

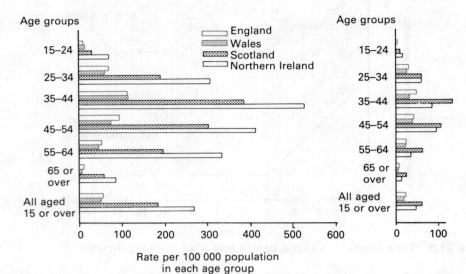

Males   **Females**

Figure 21.4 Hospital admissions due to alcohol-related mental illnesses and alcohol misuse in 1981 (source: Department of Health and Social Security; Scottish Health Service; Common Services Agency; Welsh Office; Department of Health and Social Services, North Ireland)

of the problem in men and women, in different age groups and in the different countries within the UK.

21.6 Further reading

The Department of Education and Science and the Department of Health and Social Security have produced a number of good short booklets on drug misuse. *Drugs, alcohol and mental health* by Alan and Vicky Cornwell (CUP, 1987) gives fuller information.

Questions

Q 12.1. Make a table like the one shown below. Put a tick in each space for which the description of a drug is true. The aspirin column has been done for you.

Q 21.2. The same man took twelve identical tests of speed of response one after the other both before and after drinking alcohol. The results are shown in Figure 21.3. Look at Figure 21.3 and answer the questions that follow.

(a) Before the man had a drink, what was the mean time it took him to respond?

(b) Both before and after having a drink, the man needed five tests to reach the quickest response of which he was capable.

(i) What do you think was happening to him during the five tests?

(ii) The minimum time he needed to respond before drinking was 0.05 second and after drinking was 0.3 second. What do you think was happening inside him during these minimum times?

Description	Aspirin	Penicillin	Heroin	Cocaine	Solvent	Alcohol
a medicine	✓					
obtained only on prescription						
pain-killer	✓					
gives a feeling of well-being						
leads to physical dependence						
leads to psychological dependence						

(c) (i) What do these tests show about the effect of alcohol?

(ii) Where in the man's body did the alcohol produce this effect and how did it get there?

Q 21.3. Look at Figure 21.4 and answer the questions that follow.

(a) Which sex had fewer hospital admissions per 100 000 population in each age group?

(b) Which age group had the greatest number of hospital admissions per 100 000 population?

(c) (i) Which country had the greatest number of hospital admissions per 100 000 population in every age group?

(ii) Which is the only age group in which this country had the greatest number of hospital admissions for females?

FERTILISATION AND FAMILY PLANNING

22.1 Introduction

Human life is a continuous cycle: children grow into adults; adults have children, who grow into adults, and so the cycle goes on.

22.2 Fertilisation

A child starts life as a single cell, a fertilised **egg cell** or **zygote**. The zygote contains in the DNA of its nucleus the instructions that are needed to make a new person. Half the instructions come in the DNA from the father in his sperm cell and half come from the mother in her egg cell.

    Fertilisation is the fusing of two **gametes**: a sperm cell and an egg cell. Gametes are described in Section 14.1. All gametes from the same parent have half the parent's DNA but no two are likely to have the same half (see Section 26.2).

    Sperm cells are formed in the two **testes** of a man. The two testes hang in the **scrotal sac** (see Figure 20.6) outside and below the body cavity, where a cooler than normal body temperature (about 35°C) is what is needed for healthy sperms to develop. Sperms are formed in vast numbers: two to three hundred million can be present in a single **ejaculation** of **semen** from the **penis**.

    Undeveloped egg cells are present in the two **ovaries** of a woman at her own birth. The ovaries lie inside the body cavity just above the groin. When she is old enough to menstruate (see Section 22.3), one egg cell is usually matured from each ovary about every other month. If two or more egg cells are matured and shed at the same time (either from the same ovary or from different ovaries) and are fertilised, the mother will have **non-identical** (**fraternal**) **twins** (or triplets, etc.). Such twins are as different from each other as any other brothers and sisters, except that they are the same age.

    When you remember that the DNA in gametes from the same parent is unlikely to be the same twice, you will realise that the DNA in any two zygotes is likely to be very different and can be regarded as certain to be different in some ways. Only if a zygote (or an embryo at an early stage) divides completely to form two or more different embryos do you get people born with the same DNA. These are what we call **identical twins** (or triplets, etc.).

    Triplets, quadruplets, quintuplets, sextuplets and septuplets (seven babies were born to a woman in Liverpool in 1987 though all of them died) are usually 263 non-identical or fraternal (from different zygotes) though it is possible for two or more of them to be identical (from the same zygote).

Before fertilisation takes place, sperms are ejaculated by the man in the **vagina** of the woman. Sperms, which are much smaller than the egg cell, must swim a long way to the egg cell. The semen, the fluid in which the sperms are ejaculated, contains food and enzymes, and the sperms themselves have **mitochondria** for the release of energy for swimming. The cells lining the female tubes also provide food. It can take sperms 48 hours to reach an egg cell.

About once a month an egg cell is shed from the surface of the ovary into the body cavity of a woman (a process called **ovulation**). The egg cell gets swept into one of the two **oviducts** (**Fallopian tubes**) in a current of fluid that flows from the ovary to the **uterus** (womb) via the oviducts. Sperms have to swim against this current to reach the egg cell. Only a few hundred of the millions of sperms that are ejaculated succeed in reaching the egg cell: those that do so are vigorous and healthy. An egg cell can be successfully fertilised only within two or three days of leaving the ovary.

Some sperms swim up the wrong oviduct where there is no egg cell. Most of those that swim up the correct oviduct will not fertilise the egg cell. None of them may do so: fertilisation may not take place. A couple who want a baby may wait months or years before the woman becomes pregnant. If fertilisation takes place, it does so about one-third of the way down the oviduct. Figure 14.6 shows a human egg cell and sperm cell just before fertilisation. The nuclei of the egg and sperm cells fuse at fertilisation, but the DNA from each parent remains separate in the nucleus of the zygote and in the nucleus of every cell of the body that develops from the zygote.

Figure 22.1 shows the female reproductive system consisting of
○ the vagina, where sperms are deposited
○ the **cervix**, the muscular neck of the uterus, which keeps the uterus sterile and has only a narrow opening through which sperms must swim

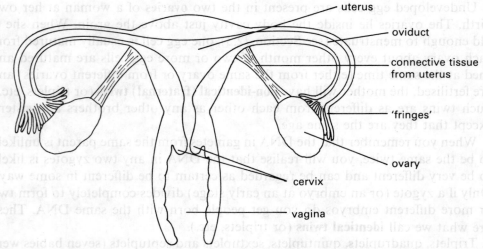

Figure 22.1 Reproductive organs in a woman (front view)

- the ovaries, where egg cells are made and stored
- the oviducts, where fertilisation takes place
- the uterus, where an **embryo**, that develops from a fertilised egg cell embeds itself in the inner lining and grows
- the wafting **fringes** of the oviducts, which help to pass the egg cell to an oviduct after it is shed
- the connective tissue, which holds each ovary in place in the body cavity

22.3 Menstrual cycle

The monthly cycle in which a woman produces an egg cell is called the **menstrual cycle**. Figure 22.2 shows the events in a 28-day cycle.

You can see from the diagram that in this 28-day cycle an egg cell is shed (ovulation) on day 14, that it can be fertilised within the next three days, and that, if not fertilised, it will die. Meanwhile the lining of the uterus develops a new blood supply ready to receive a fertilised egg if one arrives. If fertilisation has not taken place, the uterus lining is shed from the body fourteen days after ovulation at a **menstrual period** (usually just called a **period**).

But cycles vary a great deal in length both in different women and at different times in the same woman: perfectly normal cycles, in which a normal egg cell is produced, can last anything from three to eight weeks or even longer. Day 1 of a menstrual cycle is always taken as the first day of a period.

The time taken for the period, the bleeding from the vagina (called **menstruation**), can vary from a few days to a week or more. The time taken for the new egg cell to mature in the ovary before it is shed also varies. The variation in the lengths

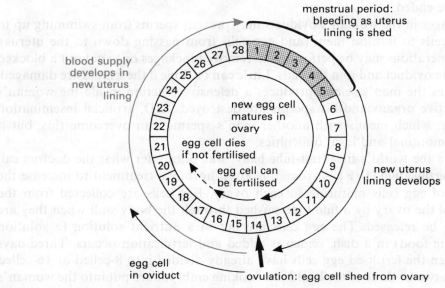

Figure 22.2   A 28-day menstrual cycle

of menstrual cycles is, therefore, not in the last fourteen days but in the earlier part of the cycle, which can take anything between a week and six weeks or even longer. This is an important fact for anyone who uses a rhythm method of family planning (see Section 22.6).

(see Section 22.6)

About one in every ten couples who want to have a baby are **infertile**. This does not mean that they cannot enjoy a normal sex life but that sex does not lead to fertilisation: the woman does not get pregnant. There are many possible reasons for a couple's infertility; some are still not understood.

Sometimes simple counselling will help, such as suggesting to the couple that they have intercourse at the time of ovulation, which is fourteen days before the woman's next period is due. Relaxation sometimes helps, as do giving up smoking and drinking and eating healthier foods. The man may be advised to wear loose pants so that his testes hang below his body and do not become overheated. One of the reasons for eating healthier foods is that, if the man is overweight, fatty tissue around his testes can also cause overheating. If the man's sperm count is low, **artificial insemination**, the use of an instrument to put the semen high into the vagina or even into the cervix, is sometimes successful. This is known as **AIH**, which means artificial insemination by husband.

The woman may not be releasing an egg cell in every menstrual cycle: injections of suitable hormones, popularly called **fertility drugs**, can cure this. Couples then run the risk of multiple births if several egg cells are released. Fertility drugs were the cause of septuplets (seven babies) born to the Liverpool woman in 1987. Multiple births can be detected very early in a pregnancy. If more than four embryos are growing, the parents and doctors may discuss whether the pregnancy should be ended.

Blockages in the woman's oviducts may prevent sperms from swimming up to the egg cells to fertilise them (and egg cells from passing down to the uterus). Simple operations may be performed to remove blockages or to cut out a blocked part of the oviduct and join the ends. Little can be done if the fringes are damaged. Sometimes the man's semen produces a defensive reaction inside the woman's reproductive organs and his sperms are destroyed. **AID**, artificial insemination by donor, which means with another man's sperms, can overcome this, but it creates emotional and legal difficulties.

In 1978 the world's first '**test-tube baby**' was born after what the doctors call *in vitro* **fertilisation** (IVF). A woman is given hormone treatment to increase the number of egg cells maturing in each ovary. Egg cells are collected from the surface of the ovary by a fine tube pushed through the body wall when they are about to be released. The egg cells are put in a nutrient solution (a solution containing food) in a dish, semen is added and fertilisation occurs. Three days later, when the fertilised egg cells have already divided into 8-celled or 16-celled embryos, two or three of the healthiest-looking embryos are put into the woman's uterus via the vagina and cervix. Three days is about the time it would have taken for an embryo to pass along the oviduct to the uterus. Several embryos are used in the hope that at least one will develop into a normal baby.

In 1987 the first test-tube baby from an anonymous donor egg cell was born to a woman in the UK. Because she was unable to produce her own egg cells, her husband's sperm was used to fertilise the donor egg cell in a laboratory dish. When the embryo was implanted in her uterus, she became pregnant and had a normal baby. Doctors have appealed to women who are being sterilised to donate egg cells for this technique.

Improvements in test-tube-baby techniques are made all the time. There are now different techniques for collecting egg cells and for placing embryos in the woman's uterus. Greater success is claimed when egg cells and sperms are placed together at the fringes end of the oviduct to allow fertilisation and the passage to the uterus to take place normally. Whatever method is used, IVF is very expensive and time-consuming, and few of the couples who try it succeed in having a baby.

22.5 World population

On 7th July 1986 the world's 5 000 000 000th person was born (according to the Population Institute of Washington). According to United Nations estimates, the world population was 4 000 000 000 in 1974 and will reach 6 000 000 000 in 1999. Figure 22.3 shows how the world population has risen over the last 9 000 years.

In 1900 the world population was less than 2 000 000 000. The population has trebled in the last 100 years. Until a few hundred years ago, natural disasters and disease kept the population fairly stable. Humans now have greater control over their environment: many diseases, such as smallpox, have completely disappeared. Population increases, like those we have seen this century, cannot be kept up:

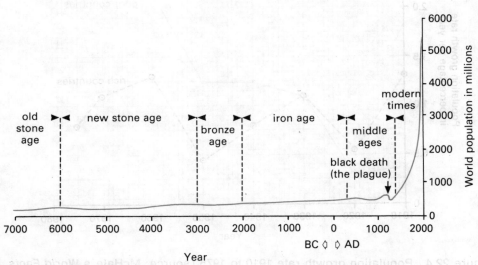

Figure 22.3   World population (source: *Population Bulletin 18*)

267

food production cannot increase indefinitely and resources such as metal ores and fossil fuels cannot be renewed. Pollution of air, land and water may become uncontrollable.

As long as the birth rate (the number of births per thousand people) in any country is higher than the death rate (the number of deaths per thousand people), the population increases. When death rates are higher than birth rates, the population decreases, In 1986 the populations of Denmark, Hungary and West Germany decreased because the death rate stayed steady while the birth rate fell below it. (We are ignoring the influence on population of immigration and emigration.)

The growth rate of the world's population is in fact decreasing. This does not mean that the number of people in the world is getting less but that the speed with which the population is rising is slowing down. Figure 22.4 shows that the population growth rate is decreasing faster in rich (industrialised) countries than in poor ones. The time will surely come when the world's population is itself decreasing.

In rich countries the growth rate is falling because the birth rate has fallen because
○ people have wanted fewer children
○ more women are going out to work and are postponing having children
○ contraception has been more widely available and more effective
○ abortion has been more widely available

The fall in the growth rate in poor countries would have been even smaller

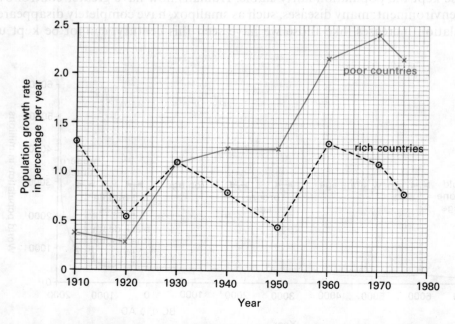

Figure 22.4 Population growth rate 1910 to 1975 (source: McHale's *World Facts and Trends*)

268

but for a rise in the death rate caused by food shortages after poor harvests, by overfishing and by the increased cost of imported food and fertilisers.

China has the world's largest population of more than 1 000 000 000 people. Its Government aims to keep the population almost steady so that by the end of the century it will be no more than 1 200 000 000. The Government's policies show its determination to reduce the population growth rate:

○ all married couples get free contraception
○ single-child families get rewards such as priority for nursery-school and primary-school places
○ the head of a single-child family gets an allowance in addition to wages
○ abortion is legal (and has been since 1957)

One way and another China makes life difficult for parents with more than one child. Parents who already have one child sometimes kill new-born babies, particularly girls, who are less likely to bring wealth into the family. Yet peasants in the countryside are defying the Government and having as many as three children. Even in Beijing, the capital, births are higher than they are supposed to be.

Other countries have been less successful in controlling population growth. The Indian Government tried giving men stereo radios for being sterilised. During the short period when India was not a democracy, the Government gave instructions that a certain number of men were to be sterilised in each district: old men who would never have had more children were sterilised because they were the easiest ones to catch.

Yet there is a simple and happy solution to the population problem. In the UK parents started having fewer children in the nineteenth century when the Government decided that all children should go to school. Instead of bringing in wages, children then became an expense: they had to be fed, clothed and housed for many years before they brought in wages. Therefore parents no longer wanted lots of children. Once this happened, parents found their own methods of contraception (even if they were unreliable): for example, women put cotton wool high up into their vaginas. It is not true, as is often suggested, that the population growth rate in the UK fell because modern contraception became available. It fell because schooling was made compulsory.

One place has shown that the population problems of the poor countries can be solved in the same way. It is Kerala, a state in south-west India. The Government of Kerala made schooling compulsory. Not only did the people of Kerala have fewer children, but they quickly began to think like people living in the rich countries. For example, parents saved to buy higher education for their children (or for their single child). Now Kerala has parents who were themselves educated as children. As a general rule, an educated population chooses to have fewer children than an uneducated one. And, as standards of health and medicine rise, people no longer have more children 'in case some of them die'.

Many Governments still think they cannot afford to introduce compulsory schooling until they have reduced their population growth. The truth is that countries are unlikely to reduce their population growth until they introduce compulsory schooling: they cannot afford not to introduce it.

Family planning is making sure couples have children only when they have decided they want them. It may mean using contraception or avoiding intercourse when fertilisation is possible. Contraception is interference with processes that could normally lead to fertilisation and pregnancy. The methods of contraception most often used are

○ the **pill** or **injections**
○ a **condom**
○ a **diaphragm**
○ an **IUD** (**intra-uterine device**)
○ **male sterilisation**
○ **female sterilisation**

Different contraceptives are shown in Figure 22.5. Preventing fertilisation by

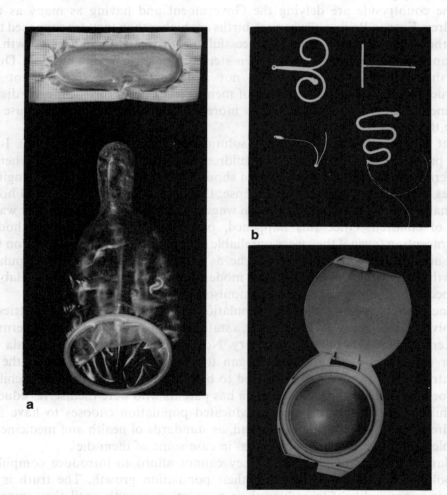

Figure 22.5 Contraceptives: (a) condoms; (b) IUDs; (c) a diaphragm in its open box

avoiding intercourse from the time of a period until after ovulation is one or other of the **rhythm methods** of family planning. It is not contraception.

(a) The pill

Different pills for women are on the market but all contain synthetic female sex hormones which prevent egg cells being shed from the ovary or prevent a fertilised egg cell embedding in the uterus lining. (The hormones are called synthetic because they are not natural ones but have been made in a laboratory or factory.) The most widely used pill is the 'combined pill' which comes in packets of 21: one is taken each day for 21 days followed by seven pill-free days: the pills contain the synthetic female hormones oestrogen and progesterone but in varying amounts. There is usually slight bleeding from the uterus for a few days in the pill-free stage. Figure 22.6 shows a typical pill cycle.

Used according to instructions, the pill is very reliable in preventing pregnancy, but some women get unpleasant side effects such as feeling sick, bloated, depressed or tired. Some pills have been linked with increased risks of heart attacks, other thromboses and cancer. The pill is unsuitable for anyone with a personal or family history of circulation problems and heart disease; it is less suitable for women over 35 than for younger women, especially if they smoke. The pill is supplied only on a doctor's prescription: pill users need to go to a doctor for regular check-ups. In many parts of the world injections of slow-release hormones are an alternative to the pill.

(b) Condom

A **condom**, usually made of thin rubber, is unrolled over a man's erect penis just before intercourse. It prevents semen containing sperms from passing into the vagina at ejaculation. It should be put on before there is any contact between penis and vagina because sperms may be present at the tip of the penis before ejaculation. The penis should be withdrawn from the vagina before an erection

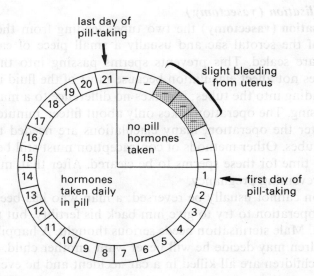

Figure 22.6   A pill cycle

271

wears off so that no semen escapes from the condom into the vagina. A condom must be used with a **spermicide**, a chemical which kills sperms and is sold as jelly, paste, foam or a pessary. Some condoms have a coating of spermicide. A condom is less reliable than the pill. Condoms are easy to get: they can be bought in chemists and other shops and from machines.

A condom reduces, though it does not remove, the risk of infection by sexually transmitted diseases, notably AIDS. This has caused an increase in the use of condoms in recent years.

(c) Diaphragm

A **diaphragm** is a thin dome of rubber which fits over a woman's cervix at the top of the vagina. It is sometimes called a cap. The right size and shape have to be prescribed by a specially trained doctor or nurse. A diaphragm has to be put in place by a woman before intercourse and should be left in place for about six hours afterwards. It must be used with a spermicide. Because the vagina may change shape or size, the fit should be checked every six months.

Like a condom, a diaphragm is a barrier method of contraception: it stops sperms swimming through the cervix. It is more reliable than a condom but less reliable than the pill, because it needs to be in exactly the right place. It has no side effects and may give some protection against cancer of the cervix.

(d) IUD (intra-uterine device)

An **IUD** is a small curved piece of plastic, usually wound with copper, in a shape such as a spiral, loop or ring. It is put in the uterus by a specially trained doctor and acts mainly by preventing a fertilised egg cell embedding in the uterus lining. In some women it results in heavy bleeding at a period and in all it increases the risk of infections in the region of the uterus. If it stays in place, it is reliable for two years or more. It is an unsuitable method for young women because infections can result in infertility: it should be used only by older women who have had all the children they want.

(e) Male sterilisation (vasectomy)

In **male sterilisation** (**vasectomy**) the two tubes leading from the testes are cut near the top of the scrotal sac and usually a small piece of each is removed. The cut ends are sealed. This prevents sperms passing into the semen. Male sterilisation does not affect ejaculation because most of the fluid in semen comes from glands leading into the tubes; it makes no difference to a man's sex life that sperms are missing. The operation takes only about fifteen minutes under a local anaesthetic. After the operation, many ejaculations are needed to clear sperms already in the tubes. Other methods of contraception must still be used for a few months to give time for these sperms to be cleared. After that, male sterilisation is reliable in preventing pregnancy.

The operation cannot usually be reversed: a man who has been sterilised can have a second operation to try to give him back his fertility, but the chances are it will not work. Male sterilisation needs serious thought. A happily married man with three children may decide he will never want another child. But what if his wife and three children are all killed in a car accident and he eventually marries again?

272 again?

(f) Female sterilisation

The usual method of **female sterilisation** is to block or cut the oviducts so that sperms cannot swim up to an egg cell and no fertilised egg cell can reach the uterus. The method is reliable and has no effect on a woman's sex life, but the operation is even more difficult to reverse than male sterilisation. Before she is sterilised, a woman needs to be sure she will never want another child. It is true, however, that, even with her oviducts blocked or cut, a woman might still have a test-tube baby.

(g) Rhythm methods

All **rhythm methods** depend on avoiding intercourse before, at and immediately after ovulation. The methods rely on being able to predict when ovulation will occur and on avoiding intercourse for some time beforehand. Sperms can live for a week or more inside the female reproductive organs and an egg cell can be fertilised in the two or three days after ovulation. The only '**safe period**' is from three days after ovulation until the start of the next period.

The time of the last period is no certain guide to the time of ovulation: all that is fairly certain is that ovulation takes place 14 days *before* the start of the *next* period. The time from the start of the last period to ovulation can vary from one week to six weeks (or even more). A rhythm method is reliable, if at all, only for a woman who has a regular menstrual cycle and can predict the date of her next period. She knows she will ovulate fourteen days before this date and will be 'safe' three days after that. In other words, she will be 'safe' eleven days before the date of her next period.

There are several rhythm methods:

o the temperature method relies on measuring a small rise in body temperature that takes place after ovulation
o the Billings or mucus method relies on detecting changes in the mucus of the cervix which take place at about the time of ovulation
o the calendar method involves predicting the safe period for intercourse from the date of the last period (or from the predicted date of the next period)
o daily urine tests are taken with a special kit
o the sympto-thermal method is a combination of some of the other methods

Rhythm methods are unreliable for women with irregular cycles and for all women after childbirth. They use nothing mechanical or chemical, which is why they are not contraception. They are used mainly by couples whose religious beliefs or principles prevent them using contraception.

(h) Unreliability

Figures for the unreliability of different methods of contraception vary greatly. So do the ways in which unreliability is measured. Table 22.1 gives figures from one survey in the UK in 100 woman-years. An unreliability of 3 in 100 woman-years means that, out of 100 sexually active women in the 15–44-year age group using that method, on average three would be pregnant by the end of one year.

Figure 22.7 gives worldwide contraception data published by the United Nations in 1984. 'Others' includes rhythm methods and withdrawal (of the penis

Table 22.1 Unreliability of contraceptive methods

Method	Pregnancy per 100 woman-years
Pill (combined pill)	less than 1.0
Condom	4
Diaphragm	2–3
IUD	2–3
Sterilisation (male and female)	less than 1.0
Rhythm methods	6–18
No contraception	80

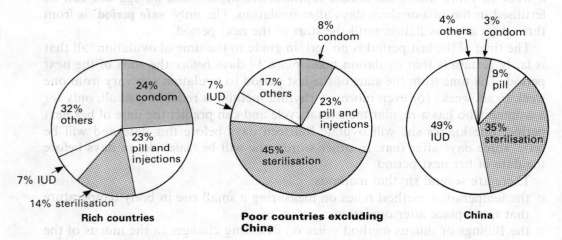

Figure 22.7   World-wide contraception (source: UN Population Division)

before ejaculation). 'Injections' are of synthetic female sex hormones: they are like the pill.

(i) The future
New pills for women, which are said to have fewer dangers and side effects, have been tried. So have a pill for men, a unisex pill, a morning-after pill for women and a vaccine for women. Better spermicides are discovered all the time.

(j) Further information
If you need help or advice, you should contact your family doctor, a family planning clinic or the Family Planning Information Service, 27–35 Mortimer Street, London W1N 7RJ (01–636 7866).

22.7 Sexually transmitted diseases

A disease passed from one person to another by any kind of sexual contact is a

sexually transmitted disease (STD). The old name for such a disease, one which is still used, is a **venereal disease (VD)**.

Syphilis and **gonorrhoea** are STDs which have been known for centuries. Both are caused by bacteria and both can be treated successfully with antibiotics. When penicillin was first used (in the 1940s), it completely cured gonorrhoea. But penicillin was used so much that the bacteria became resistant to it (by developing an enzyme that breaks down penicillin). Other antibiotics are now used. Tracing the sexual contacts of infected people and treating them have reduced the spread of these diseases.

Genital herpes is caused by a virus and, like all viruses, cannot be treated with antibiotics. The virus stays dormant inside body cells for long periods. When it becomes active, it causes in the genital region a painful sore which oozes a colourless fluid. Like all the STDs, it infects both men and women and is passed on by sexual contact.

The worst of the STDs is **AIDS**, which may be caused by a virus called the **human immunodeficiency virus (HIV)**. AIDS stands for **Acquired Immune Deficiency Syndrome**: *acquired* because it is caught from someone, *immune deficiency* because it reduces the body's ability to overcome certain illnesses, and *syndrome* because infected people tend to get the same illnesses and symptoms. Research to find a cure for AIDS, and a vaccine to prevent it, is world-wide. White blood cells are attacked, which is why AIDS reduces a person's resistance to other infections (see Section 14.1). People with AIDS die of other diseases such as pneumonia.

Not all those who are infected with HIV get AIDS: some have a brief illness like 'flu; others get no symptoms at all for years. No one knows yet what proportion of people with the virus are likely to get AIDS. At present there is no cure. Most people who get AIDS die within two years. The Department of Health and Social Security's figures for the UK from 1982 to the end of July 1988 are:

total number of AIDS cases	1669 (1617 men and 52 women)
total number of deaths	916

At least 40 000 people in the UK were thought, at that time, to have been infected with the virus. There is uncertainty about the disease because research on it began only in the 1980s.

AIDS is passed from one person to another in three main ways:
○ by sexual contact, both heterosexual and homosexual
○ by infected blood, for example, on shared needles among drug users
○ from a woman to her fetus
The risk to most people is in sexual contact. The more sexual partners, the greater the risk of getting the virus. The use of condoms (see Section 22.6) is known to reduce the risk of infections by all STDs, including AIDS, but is not foolproof. Condoms are designed as a barrier to sperms, which are about 3 μm wide. *Gonococcus*, the bacterium that causes gonorrhoea, is only 0.8 μm wide while the herpes virus and the HIV are only 0.1 μm wide. (1 μm is one-thousandth of a millimetre.) Oil-based lubricants should not be used with condoms because they can damage rubber.

Further information about all STDs is easily obtained from special STD Clinics   275 listed in the phone book under Sexually Transmitted Disease or Venereal Disease. The Family Planning Information Service (01–636 7866) will also help. There is a blood test which will show if a person has been infected with HIV but no test can tell if a person infected with HIV will go on to develop AIDS.

Q 22.1. Make a table like the one shown below but larger. Give the advantages and disadvantages of the methods of family planning which are listed in the left-hand column.

Method of family planning	Advantages	Disadvantages
Pill (combined pill) Diaphragm Condom IUD Male sterilisation Female sterilisation Rhythm methods		

Q 22.2. What do the following letters stand for: AID, AIDS, AIH, HIV, IUD, IVF and STD?

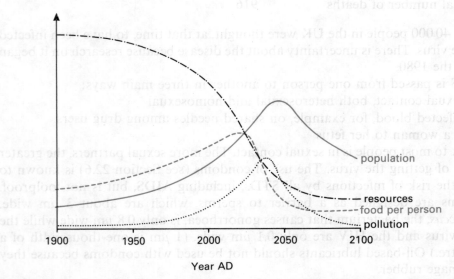

Figure 22.8 World trends and predictions (source: D. H. Meadows, New York)

Q 22.3. Assume that: a menstrual cycle lasts 25 days; ovulation takes place on day 11; egg cells can be fertilised up to three days after ovulation; sperms can live in the female reproductive organs for up to seven days. When must intercourse not take place if a pregnancy is to be avoided?

Q 22.4. Figure 22.8 shows world trends in population, food per person, pollution and resources since 1900 together with computer predictions for the future by a team in New York in 1972.

(a) Describe the predicted trends from 1972 to the present for (i) population, (ii) food per person, (iii) pollution and (iv) resources.

(b) What could be done to prevent food per person falling to such a low level after the year 2050?

PREGNANCY

23.1 Introduction

Doctors working on *in vitro* fertilisation (test-tube babies) can see how, in the first few days after an egg cell is fertilised in the oviduct, it divides into two, four, eight and sixteen cells. In a normal pregnancy this is when it is travelling down the oviduct to the uterus.

23.2 Implantation

Pregnancy (the **gestation period**) is the time from fertilisation until birth. At the start of a normal pregnancy the dividing cells are swept down the oviduct in a fluid current. Meanwhile the new lining of the uterus is thickening and developing an improved blood supply under the control of hormones released from the ovary after ovulation. The new lining develops whether an egg cell is fertilised or not. If the egg cell is not fertilised (and most egg cells are not) the new lining of the uterus breaks down and is shed at menstruation fourteen days after ovulation (see Section 22.3).

Figure 23.1 shows that the ball of cells, into which the fertilised egg cell soon develops, arrives in the uterus about five days after ovulation. Figure 23.1 also

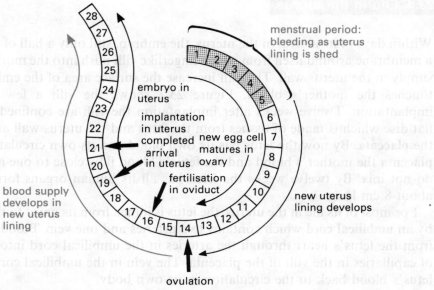

Figure 23.1 Early pregnancy

shows that **implantation**, the embedding of the fertilised egg cell in the lining of the uterus, is completed after two more days. From now until the end of the pregnancy, the developing baby gets all its food and oxygen by diffusion from its mother's blood supply. After only a few days, when it is little more than a group of cells, it makes and releases hormones that will stop menstruation and prevent the uterus lining from breaking down and being shed.

23.3 Terms

A lot of confusion exists about the names for a developing baby. The term *zygote* is used for any cell that is formed by fertilisation, the fusion of two gametes. The human zygote is formed when egg cell and sperm fuse. The term *embryo* is used for any young organism in the early stages of its life. Biologists talk about plant embryos inside seeds. But doctors use the term *embryo* only for the developing baby from the time the zygote divides into two cells until the mass of cells begins to look like a human baby. This happens remarkably early in pregnancy, when the developing baby is only about six weeds old and 4 cm long. From then until birth doctors call it a *fetus*.

The term *pre-embryo* may be used for the first fourteen days after the division of the zygote because most of the cells at this stage will form tissues surrounding the true embryo rather than the true embryo itself.

Another confusion exists about the time pregnancy lasts. The average time a baby develops inside its mother is 266 days, which is 38 weeks or about one week less than nine calendar months. A pregnancy can last several weeks more or less than this average time. In the following Sections the figures refer to the time after fertilisation.

23.4 Life in the uterus

Within days of its arrival in the uterus, the embryo, still only a ball of cells, forms a membrane around itself from which fingerlike **villi** push into the mother's blood supply in the uterus wall. The villi increase the surface area of the embryo which touches the mother's blood. Figure 23.2 shows the villi a few days after implantation. Twelve weeks after implantation the villi are confined to a large flat disc which is made of tissues from the fetus and the uterus wall and is called the **placenta**. By now the fetus has its own blood and its own circulation. At the placenta the mother's blood and the fetus's blood flow close to one another but do not mix. By twelve weeks the fetus has all its human organs formed and is about 8 cm long.

For most of its life in the uterus, the fetus is joined from its navel to the placenta by an **umbilical cord** which contains two arteries and one vein. Blood is pumped from the fetus's heart through the arteries in the umbilical cord into a network of capillaries in the villi of the placenta. The vein in the umbilical cord takes the fetus's blood back to the circulation in its own body.

280   The fetus is surrounded by **amniotic fluid**, which takes all pressures off it, allows

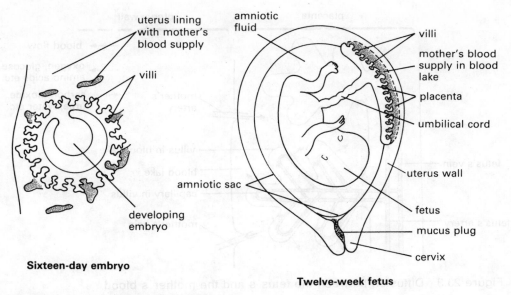

Sixteen-day embryo

Twelve-week fetus

Figure 23.2   Development of placenta

it to grow, makes it easier for it to move about and protects it from any knocks it might get through the uterus and abdominal walls of the mother. The fluid is kept in place around the fetus by a membrane, called the **amniotic sac**, which lines the uterus wall.

The mother's blood surrounds the placental villi in what are called **blood lakes**. Blood is brought to the mother's side of the placenta by arteries in the uterus wall and leaves the placenta by veins in the uterus wall. There is a continuous flow of the mother's blood through the blood lakes: it brings oxygen and food (for example glucose and amino acids) for the fetus's blood and takes away waste products (for example carbon dioxide, water and urea) from the fetus's blood.

The mother's blood and the fetus's blood do not mix: in the placenta substances **diffuse** from one circulation to the other because the concentration is higher on one side than the other, providing a concentration gradient. Figure 23.3 shows events in the placenta in greater detail.

Why is there a concentration gradient at the placenta? Blood coming from the fetus has had its oxygen and food used up by the growing cells, while the blood from the mother has come from her heart and before that from her lungs and body (see Section 19.3) and has a high concentration of oxygen and food. Blood from the fetus is carrying carbon dioxide and water (from respiration in the cells of the fetus) and urea (from any excess amino acids), while blood from the mother gave up carbon dioxide at her lungs and got rid of some urea and water when it last passed through her kidneys: it will have a lower concentration of these substances. Thus there are concentration gradients between mother and fetus in oxygen and food in one direction and in carbon dioxide, water and urea in the opposite direction.

Figure 23.4 shows a fetus in the uterus at five months through the pregnancy. The fetus now takes up much more space in the uterus and its mother can feel   281

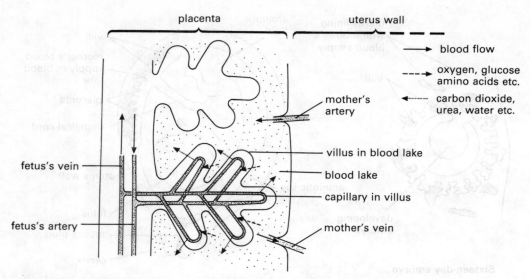

Figure 23.3   Diffusion between the fetus's and the mother's blood

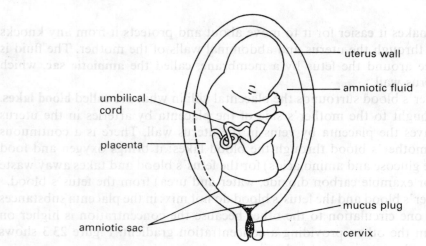

Figure 23.4   Fetus at five months

it to grow, makes it easier for it to move about and protects it from any knocks it might get through the mother's abdominal wall, or the mother. The fluid is kept in place around the fetus by a membrane called the amniotic sac, which lines the uterus wall.

The mother's blood surrounds the villi in fluid-filled blood lakes. Blood is brought to the mother's side of the placenta by arteries in the uterus wall and leaves the placenta by veins in the uterus wall. There is a continuous flow of the mother's blood through the placenta, bringing oxygen and food (for example glucose and amino acids) for the fetus's blood, and takes away waste products (for example carbon dioxide and urea) from the fetus's blood.

The mother's blood and the fetus's blood do not mix in the placenta. Substances diffuse from one circulation to the other because their concentration is higher on one side than the other. Villus on the concentration gradient. Figure 23.3 shows events in the placenta in greater detail.

Why is there a concentration gradient at the placenta? Blood coming from the fetus has had its oxygen and food used up by the growing cells, while the blood going to the body (see Section 13.3) and has a high concentration of oxygen from the fetus is carrying carbon dioxide and water from respiration in the cells of the fetus and from any excess amino acids), while the blood coming from gave up carbon dioxide at her lungs and got rid of some from the mother. It has just passed through her kidneys. It will have a lower concentration of these

it moving about inside her. Before it is born, a fetus usually turns to come out head downwards.

23.5 Ultrasound scans

Most women in the UK live within reach of a hospital with scanning equipment and can therefore have an **ultrasound scan** when they are pregnant. The woman lies on her back under the scanner and her abdomen is oiled so that the scanner can be moved across her skin without pain. The scan uses sound waves to build up a picture of the fetus in the uterus. The picture is shown on a television monitor.

Ultrasound scanning, which is harmless, is usually done when the fetus is about sixteen weeks old. It is done to check that the fetus is growing normally and that it is the age the doctors have calculated. If a woman has irregular periods, it is easy to make a mistake in calculating when she became pregnant. A scan may show that she will have her baby a month earlier or a month later than she had been told. Only if a penis can be seen in the scan is the sex of the baby certain. If no penis can be seen, the baby may be either a girl or a boy. The techniques of amniocentesis and chorionic villus sampling (see Section 23.9) make it possible to tell the sex of a fetus with certainty.

Figure 23.5 is a photograph of a scan of a nineteen-week fetus. The lines across the photograph are produced by the scanning technique. The fetus is lying slightly curled up with its legs drawn up to its body, like the fetus in Figure 23.4. Its skull and vertebral column are the lighter curves at the top of the picture; the umbilical cord is the lighter wavy line below halfway on the left. The mother's bladder is the very dark patch on the right: it has to be full of urine to get a clear scan.

23.6 Abortion

Doctors believe that more than half the egg cells that are fertilised do not embed properly in the uterus lining and are then shed, together with the uterus lining, at an early stage in pregnancy. A woman's period may not even be late when this happens, when she will never know an egg cell was fertilised. This is a very early **abortion**.

The term *abortion* is used for any loss of an embryo or fetus that does not reach an age at which it can survive out of the uterus. An abortion may be **spontaneous**, that is accidental, and occur at any time throughout the pregnancy

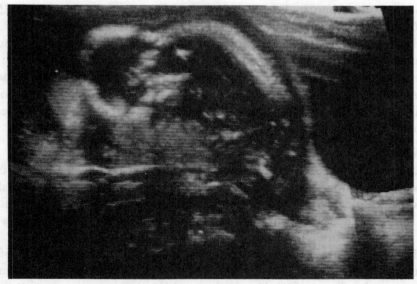

Figure 23.5   Scan of a nineteen-week fetus

until the fetus can survive. There is a tendency for abortions to occur either in the early stages of pregnancy or at what would have been the time of a period. Spontaneous abortions occur for all manner of reasons: the most common reason is that something is wrong with the developing embryo or fetus; another common reason is the ill health of the mother. Spontaneous abortions used to be called **miscarriages** (and still often are by the general public).

An **induced abortion** occurs when there has been a deliberate act to end the pregnancy. An induced abortion is legal in the UK provided that two doctors have agreed that it is necessary and that it is carried out in an approved hospital or clinic before the fetus is a certain age. There are many reasons why an induced abortion may be necessary. For example, it is possible for an embryo to develop in the wrong place: in the oviduct or even in the body cavity. This is an **ectopic** pregnancy, which may need an immediate abortion to remove the embryo. The usual methods of inducing an abortion are

- ○ **vacuum aspiration**, in which amniotic fluid and the small embryo (or fetus) are drawn out through the cervix by suction
- ○ **dilatation and curettage** (**D and C**), in which the cervix is widened and the contents of the uterus are emptied out
- ○ **prostaglandin pessaries**, which are put in the vagina and which contain hormones that induce **labour** (see Section 24.2)

23.7 Hazards for the embryo and fetus

Every pregnant woman hopes that her baby will be born without any abnormalities. About two babies in a hundred are born with an abnormality of some kind. Abnormalities range from a serious heart defect to a sixth toe (which is easily removed). Some abnormalities are inherited (see Sections 24.3, 25.3, 25.5, 26.4 and 26.6); others develop while the baby is in the uterus.

In the first four weeks of the embryo's development, when it is little more than a group of cells, any damage at all to it will probably cause it to die and be aborted. Damage between four and seven weeks after fertilisation, when the baby's organs are forming, may not cause death but will probably cause abnormality. Damage is less serious after twelve weeks, when the baby's organs are formed and just get bigger.

(a) Rubella
German measles or **rubella** is caused by a virus. It is one of the mildest common illnesses in childhood causing only a headache, a stiff neck, a faint rash which lasts for a couple of days and a slight rise in temperature. Even in an adult it is not a serious illness. But it may be damaging to an embryo or fetus if a woman gets it in the first three months of pregnancy.

In 1941 an Australian eye surgeon first noticed that babies more often had cataracts in their eyes if their mothers had had rubella during pregnancy. (A cataract is a cloudiness of the lens in the eye.) Further investigations at a London hospital showed that babies of mothers who had had rubella, by comparison with babies whose mothers had not had rubella, were

      7 times more likely to have abnormalities
    14 times more likely to have heart abnormalities
    78 times more likely to have cataracts
    40 times more likely to be deaf

All abnormalities were more likely the earlier the mother had rubella in the pregnancy. Rubella in the mother after the fetus is twelve weeks old is unlikely to cause any abnormality, though a few cases of deafness have been reported in babies whose mothers had rubella in the fourth month of pregnancy.

It seems that the rubella virus, which is tiny, can pass across the placenta and damage certain organs as they develop. The chicken pox and mumps viruses are among others that can also damage an embryo's developing organs. After you have had these diseases, your body has antibodies against them and you are unlikely to get them again. Because rubella is likely to harm an embryo if caught by a woman during early pregnancy, and because rubella is not serious at any other time, it is desirable for all girls to catch it while they are young. If you are a girl and have not had rubella, you should seek out any friends who have it, not avoid them. If you do not succeed in catching it while you are young, you should see a doctor and have an injection against it before you intend to become pregnant.

(b) Smoking

The first major study of the effects of smoking during pregnancy was not carried out until 1957: it showed that babies whose mothers smoked during pregnancy tended to weigh less than other babies. Dozens more studies have since confirmed this. Researchers tried to make sure that low birth weight was not due to other possible causes such as low incomes and diet. Studies have also shown that, the more cigarettes smoked during pregnancy, the lower the average birth weight and the greater the proportion of mothers who have babies with a low birth weight (below 2500 g). Table 23.1 shows data from a study in 1980.

Other studies have shown that, if women who normally smoke give up smoking during pregnancy, their babies weigh more (on average) than the babies of women who smoke – and often more than the babies of women who have never smoked!

Table 23.1   Birth weights and smoking during pregnancy

	Number of cigarettes smoked per day			
	0	1–10	11–20	20+
Average birth weight in g	3399	3272	3185	3128
% birth weight below 2500 g	4.8	8.0	9.0	13.4

Is it fair to force your baby to smoke cigarettes?

This is what happens if you smoke when you're pregnant.

Every time you inhale you fill your lungs with nicotine and carbon monoxide.

Your blood carries these impurities through the umbilical cord into your baby's bloodstream.

Smoking can restrict your baby's normal growth inside the womb.
It can make him underdeveloped and underweight at birth.
Which, in turn, can make him vulnerable to illness in the first delicate weeks of his life.
It can even kill him.
Last year, in Britain alone, over 1,000 babies might not have died if their mothers had given up smoking when they were pregnant.

If you give up smoking when you're pregnant your baby will be as healthy as if you'd never smoked.
Turn over and find out how to stop.

Figure 23.6   Health Education Council leaflet

The Health Education Council (no longer in existence) issued warning leaflets, like the one shown in Figure 23.6, to discourage pregnant women from smoking.

The babies of mothers who smoked during pregnancy are more likely to die during the **perinatal** period. The perinatal period is from shortly before until shortly after birth: it is usually taken to be from the 28th week of pregnancy until the time when the baby is a month old. The lower average birth weight of smokers' babies is likely to be one reason why they are more likely to die.

We do not know for certain why the babies of pregnant smokers have a lower average birth weight. But a pregnant woman's red blood cells can carry less oxygen if they have picked up carbon monoxide from the smoke in cigarettes (see Section 18.5). If there is less oxygen in the mother's blood, the fetus gets less oxygen: the fetus needs oxygen for respiration to release energy for growth. There are other possible explanations, for example, that the nicotine in cigarette smoke, which stimulates the release of adrenaline into the blood, constricts some of the arteries in the fetus and deprives its tissues of some blood. Nicotine is also known to damage DNA in the placenta.

(c) Alcohol

Alcohol easily passes through the placenta into the fetus's circulation. The concentration in the fetus's blood will be the same as in the mother's because there is free diffusion from the mother's blood to the blood in the fetus's capillaries. If the mother is 'over the limit', so is the fetus. An occasional small drink may not matter, but regular drinking during pregnancy reduces the baby's birth weight. There are data to show that for every 10 g alcohol drunk daily by the mother
○ the average growth rate of the fetus is reduced by 1 per cent
○ the risk of abnormalities rises by 1.7 per cent
The point of these data is that 10 g of alcohol is not a lot: it is just over half a pint of beer or one glass of table wine.

If pregnant women are heavy drinkers, there is a risk that their fetuses will have small heads, small eyes, badly developed face bones and heart abnormalities and that they will be mentally retarded.

(d) Drugs

All drugs, whether prescribed by doctors or taken for kicks, interfere with processes in the body (see Section 21.5). A doctor should prescribe a drug only if its likely good effects outweigh its possible harmful effects. Drugs may have side-effects which doctors do not yet know about. A good rule is never to take any drug if you think you can manage without it. This is all the more important for a pregnant woman: drugs may not harm her but they may harm the fetus.

It used to be thought that the placenta was a good barrier between mother and fetus and that many substances in the mother's blood would not pass into the fetus's circulation. More and more substances are now known to cross the placenta. The tragedy of the thalidomide babies has made the public more aware of the dangers and has made doctors more cautious about what they prescribe for pregnant women.

Thalidomide was first made in West Germany in 1954. It was prescribed as a sleeping tablet because overdoses did little harm. In 1958 a baby was born in West Germany without arms or legs. No one connected this with thalidomide because such events, though rare, had occurred before. In November 1961, at a medical conference, 34 babies with abnormal arms and legs were described. Dr Lenz of Hamburg suggested that there might be a link between them and thalidomide taken during pregnancy. Professor Smithells of Liverpool had noticed five babies with similar abnormalies of the limbs. He checked back and found   287 that all their mothers had taken thalidomide early in their pregnancies. Within nine days the drug was withdrawn.

A great deal is now know about thalidomide. Eight hundred affected children were born alive in England, 2800 in West Germany. Thalidomide affected children whose mothers had taken the drug between the 37th and 50th day after the start of the last menstrual period (when they were 5–7 weeks pregnant), even if they had taken only one dose. Long bones of the limbs were often missing, deformed hands and feet grew on the trunk, outer ears did not form properly and there were often internal abnormalities of the heart, kidneys, gall bladder and appendix. Brain damage was rare and there seems to be no damage to the DNA: none of the abnormalities should be inherited.

23.8 The antenatal clinic

Times have changed since doctors kept their patients ignorant of what was happening to them. One of the recognised aims of **antenatal clinics** is to give pregnant women information about themselves and their developing babies. Run by local health authorities and staffed by specially trained doctors and midwives, antenatal clinics check the health of pregnant women and, as far as possible, the health of their fetuses by questioning and by examination. Records are kept.

Pregnant women are given information freely and the events of pregnancy and labour are explained to them. Many health authorities arrange classes and discussion groups in which pregnant women are encouraged to ask the questions that are worrying them. They are also encouraged to see their dentists: dental care is free in the UK for pregnant women. Some general practitioners who have had special training hold their own antenatal clinics. Some antenatal clinics are held at local hospitals.

The first visit to an antenatal clinic usually takes place when a woman is twelve to thirteen weeks pregnant. A nurse records details of each woman who attends, including her medical history, her weight and height, and often too her shoe size. A woman's height and shoe size give a clue to the size of her pelvis, which needs to be a certain width to allow the easy passage of her baby's head at birth.

A doctor or midwife takes her blood pressure and checks the condition of her heart, lungs and teeth. Feeling the outside of the abdomen gives the doctor an idea of the size of the uterus and how far the pregnancy has got. Some doctors make an internal examination through the vagina; others do not do so until later, if at all. Women may give a blood sample and a urine sample at the first visit. The first visit at twelve to thirteen weeks is important because the progress of the pregnancy can be assessed only against the readings taken then (for weight, blood pressure, etc.).

(a) Weight

A woman's weight is usually recorded at each visit. This allows her weight gain (or occasionally her weight loss) to be calculated each time. A gain in weight is of course normal as the fetus develops. But a large weight gain suggests something may be wrong. If the weight gain is due to storage of water in the woman's tissues, which will usually show up as puffiness in her ankles and fingers, and if she also has high blood pressure and protein in her urine, her condition needs immediate treatment (see below).

The more weight a woman puts on during pregnancy, the more she needs to lose afterwards to regain her figure. A weight gain of about 12 kg (26 lbs) through the entire pregnancy is a healthy maximum. Encouraged to 'eat for two', pregnant women used to eat too much and get overweight, which made everything an effort. There is no need for a pregnant woman to eat a lot 'to be on the safe side'. Even if she eats too little, her developing baby will get all the food it needs: it is the woman's own body that will go short. Table 16.1 shows how much extra energy food, protein, vitamin D and calcium is needed by a woman who is four months' pregnant.

(b) Blood pressure

A woman's blood pressure is recorded at each visit. Blood pressure varies among perfectly normal people (see Section 19.4), but raised blood pressure during pregnancy, which a woman may not even notice, can damage the placenta and therefore the fetus. Raised blood pressure in a pregnant woman should be treated as soon as it is discovered; if her blood pressure has gone up a lot, she may need a stay in hospital.

(c) Blood sampling

A blood sample is sent away for laboratory testing. Technicians determine the woman's blood groups, in particular her ABO group and her Rhesus factor. If she is Rhesus negative and the fetus is Rhesus positive, her antibodies may damage the fetus's blood: the fetus's blood will be examined before and at the birth. The technicians also discover whether she is **anaemic** (lacking red blood cells and haemoglobin). Anaemia, which would deprive the fetus of oxygen, can usually be treated successfully with tablets of iron or iron and folic acid. The technicians can also test whether she is immune to rubella and whether she has certain serious diseases such as syphilis and AIDS.

(d) Urine sampling

At every visit to the clinic a woman's urine is tested for the presence of glucose and protein. The presence of glucose is serious only if a blood test after several hours without food shows that the woman is **diabetic**. There is a slightly higher risk of abnormalities (6 per cent instead of 2 per cent) amont the babies of diabetic women: some constriction of arteries leading to the placenta may reduce the supply of oxygen to the fetus. Diabetic women need to be identified early in pregnancy so that they may be given special treatment which will reduce risks to their babies.

Protein in the urine, together with high blood pressure and puffiness of the tissues (because water is being stored in them), is a sign of **pre-eclampsia** (formerly called toxaemia), a form of poisoning the cause of which is unknown. With all three symptoms present, a pregnant woman needs immediate treatment in hospital if the fetus is not to be harmed. Protein in the urine is also a common sign of a urinary infection. Such infections need treatment.

289

In the early 1970s it was learned how to remove amniotic fluid which contained cells from the fetus and to get these cells to grow and multiply in a nutrient solution. This technique, known as **amniocentesis**, made it possible to identify abnormalities such as **Down's syndrome** (see Section 26.4). Amniocentesis is now used to identify a large number of different inherited abnormalities not only from the cells but also from chemical abnormalities in the amniotic fluid. Figure 23.7 shows the main stages in the technique.

Amniocentesis is usually done during the sixteenth week of pregnancy and it takes several weeks before the results are known. If severe fetal abnormalities are discovered, an induced abortion is recommended to the parents. A baby born with **spina bifida** has part of its spinal cord exposed and will have to have many operations. Cases of spina bifida are a quarter what they were in the early 1970s because it can be identified early in pregnancy and the fetus can be aborted.

An abortion at any time is distressing for parents. The later in the pregnancy that it takes place, the more distressing it is. By twenty weeks, when the results of amniocentesis are known, parents are emotionally involved with the new baby and are beginning to plan for its arrival. The fact that by this time the mother can feel the baby moving about inside her makes an abortion more distressing.

A new technique, **chorionic villus sampling** (CVS), can be used as early as nine weeks into the pregnancy and its results are known within a few days. CVS removes tissue from the villi at the edge of the placenta. It is because many more cells can be taken than in amniocentestis that results are known within a few days. If an abortion is needed, it can be carried out much earlier with less distress to the parents. In detecting some but not all abnormalities in the fetus, CVS can replace amniocentesis.

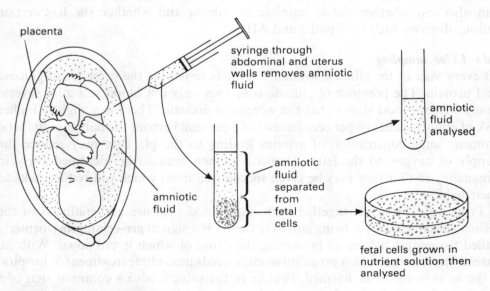

placenta

syringe through abdominal and uterus walls removes amniotic fluid

amniotic fluid analysed

amniotic fluid

amniotic fluid separated from fetal cells

fetal cells grown in nutrient solution then analysed

Figure 23.7 Amniocentesis

Q 23.1.

(a) What are the meanings of *implantation* and *abortion*?

(b) A human baby has been a zygote, an embryo and a fetus. Define each of these three stages.

Q 23.2. How may a developing baby in the uterus be harmed if the mother (a) smokes throughout the pregnancy, (b) drinks alcohol regularly throughout the pregnancy and (c) catches rubella (German measles) in the first three months of the pregnancy?

Q 23.3. Make a table like the one shown below but larger. Explain why the measurements and tests listed in the left hand column are carried out on a pregnant woman at an antenatal clinic.

Measurements and tests	Why carried out
Weight Height Blood pressure Blood sampling Urine sampling	

Q 23.4. Table 23.2 shows the results of four different studies comparing mean (average) birth weights and the smoking habits of the mothers.

Table 23.2 Mean birth weights and mothers' smoking habits

Year of study	Birth weight in g		
	Smokers	Smokers who gave up during pregnancy	Non-smokers
1959	3119	3312	3299
1966 boys	3192	3612	3472
girls	3052	3472	3360
1972	3171	3305	3316
1977	3265	3342	3451

(a) What do these figures show about the average birth weights of boys and girls?

(b) What do these figures show about the average birth weights of babies whose mothers smoked throughout pregnancy, gave up smoking during pregnancy and were non-smokers before and during pregnancy?

BIRTH TO OLD AGE

Independent life begins at birth. The baby grows into a child, an adolescent and an adult who ages and finally dies.

24.2 Birth

A baby is born, on average, after 38 weeks in the uterus. Most births in the UK take place in hospital; in other parts of the world births take place at home and birth is thought of as a more natural process. Hospital is the best place for births if there are medical complications. Usually birth is straightforward but it is not always possible to predict which births will have complications and need hospital care.

During the last few weeks of pregnancy, muscles in the uterus wall contract at intervals more and more strongly. Most women do not notice these contractions until they suddenly become powerful and more regular: this is the start of **labour**, the process of birth. The muscles squeeze the amniotic fluid in the uterus until the amniotic sac bursts (this is called 'breaking the waters'). Sometimes leakage from the vagina of the amniotic fluid shows that labour has begun before the woman notices her contractions.

Before this time the baby has usually turned in the uterus so that it is now lying head down with its head fitting neatly into the **pelvis** (a Latin word that means *basin* and describes the shape of the pelvis). The baby will then come out head first. If the baby has not turned, and if a doctor or midwife cannot turn it, it will come out feet or bottom first.

As contractions of the uterus become more powerful and more regular, the muscles surrounding the cervix relax so that, as the baby is pushed down, the opening in the cervix gradually widens and the cervix is stretched around the baby's head (if it is coming head first). The contractions of the uterus now change direction and, with the help of contractions of the muscles in the abdominal wall, force the baby through the vagina towards the outside world.

The baby is still attached to the placenta by the umbilical cord, which continues to supply it with oxygen throughout labour. At birth the baby takes its first breath of air and its circulation changes: more blood goes to the lungs and no more goes to the placenta.

A midwife clamps and cuts the umbilical cord, which shrivels during the next few days to form the **navel**. A short while later, the uterus contracts to push out the placenta and amniotic sac, called the **afterbirth**. The midwife may help to pull   293 out the afterbirth at the time of the contraction. This is the last stage of labour. The average time of labour for a first baby is thirteen hours; for a second baby and later babies it is eight hours. Figure 24.1 shows stages in the process of labour.

At antenatal clinics there is training in relaxation and in techniques of muscles contraction of the abdomen to shorten labour and make it less painful. Labour is hard work for the mother, which is how it gets its name. Drugs can be given to make it less painful.

(a) Induction

If the birth of the baby is much later than expected, or if there is a risk to the health of the mother or the baby from, for example, high blood pressure, birth may be started artificially. This is **induction** of labour. Usually a hormone

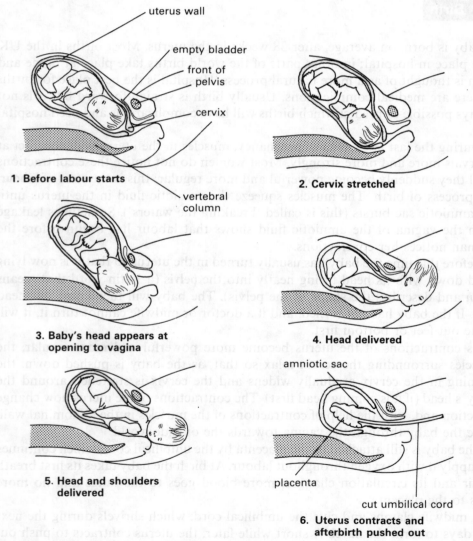

1. **Before labour starts**

2. **Cervix stretched**

3. **Baby's head appears at opening to vagina**

4. **Head delivered**

5. **Head and shoulders delivered**

6. **Uterus contracts and afterbirth pushed out**

294   Figure 24.1   Labour

(oxytocin) is given to the mother either in a slow drip of fluid into a vein or by injection. The hormone, together with the release of some amniotic fluid by the doctor or midwife, is nearly always successful in starting contractions.

(b) Caesarean section

If induction has not resulted in labour, if the baby is distressed, or if birth through the cervix and vagina would be dangerous for the mother or the baby (perhaps because the pelvis is too narrow or the placenta is lying across the cervix), a **Caesarean section** is carried out. In this operation the abdominal wall and uterus are cut at the front, the baby is lifted out and the uterus and abdominal wall are stitched up again. It is said the operation is called a Caesarean because Julius Caesar was born this way. It is more likely to be because Caesar restated the ancient law that a woman dying in labour must be cut open in the hope of saving the baby.

24.3 Post-natal screening

A few days after the baby's birth (before it leaves hospital if it has been born there) it is examined and given tests to find out if everything about it is normal. One of the tests it is given is for **PKU**.

(a) Testing for PKU

Phenylketonuria (PKU) is an inherited disease (see Section 25.3) in which the body has no enzyme for dealing with a particular amino acid (phenylalanine). The body cannot deaminate this amino acid as it can all others (see Section 20.5). If it accumulates in the body, it affects the nervous system and eventually causes brain damage and mental retardation.

A test for PKU is now done on a sample of the baby's blood or urine soon after its birth: a simple test can be done with paper on urine in a nappy. If the baby has the disease, a diet for life, free of proteins that contain this amino acid, will prevent brain damage.

(b) General examination

During a general medical examination the baby's eyes, ears, heart and lungs are checked. Its **reflex actions** are tested. Its responses should include: a **gripping reflex** when something is put in each hand; a **sucking reflex** when its lip or skin near its mouth is touched; a **startle reflex** (in which it throws out its arms and then flexes them) when it is sat up and allowed to start falling backwards; a **stepping reflex** when it is held up by its body with its feet on something. Hip dislocations are looked for: these can be easily treated in a baby but lead to serious disability later if they are not.

24.4 Feeding

As soon as the placenta separates from the uterus wall, the hormones that it has released to keep the pregnancy going no longer pass into the mother's blood.

The separation of the placenta sets up a new hormone system by which milk is produced from glands in the mother's breasts. Her breasts have already been enlarged by the development of these glands and ducts early in pregnancy.

The first secretion from the breasts, soon after the birth or even before it, is a substance called **colostrum**. It contains white blood cells and a higher proportion of the antibodies (see Section 19.2) usually present in milk. The composition of colostrum is something between that of milk and blood plasma. Colostrum does not need much digestion and is the right food for a baby which has not used its own digestive system in the uterus. Table 24.1 shows the main substances (apart from water) in colostrum, human milk, cows' milk and blood plasma.

Not only is colostrum obviously of value to the baby but the sucking of the baby at the nipple is of value to the mother. It encourages the release of more hormones which shrink the blood vessels and muscle tissue in the uterus wall and bring the uterus back more quickly to the size it was before pregnancy. It has to return from a final mass of about 1 kg (1000 g) to its usual mass of about 50 g.

After a few days milk takes the place of the colostrum. The glands in the breasts (**mammary glands**) make the milk by extracting what is needed from the blood as it passes in capillaries through the breast tissues. (Section 20.2 describes how the sweat glands in the skin also extract what is needed from blood capillaries.) Milk is made and stored in the mammary glands in the breasts: when the baby sucks at the nipple, the milk squirts into the baby's mouth in fine streams.

Many mothers have breast pains during the first few days of breast-feeding. Usually they do not last. When babies cry, mothers worry that they are not getting enough milk. Almost all babies cry: it is not often due to underfeeding. The more often the baby is breast-fed, the more milk the mother produces: a baby fed approximately every three hours gets more milk than one fed every four hours. Bottle-feeding a baby part of the time is not a way of making sure it gets enough to eat: the result is that the mother produces less milk.

Breast-feeding is better than bottle-feeding. Human milk is the natural food for babies. Some babies are allergic to cow's milk (though it may not be realised at the time and may create food-allergy problems in later life). Other advantages of breast-feeding are:

o the milk is sterile and is the right temperature and the right formula

o the colostrum contains white blood cells, and both colostrum and milk contain

Table 24.1 Contents of colostrum, human and cow's milk and blood plasma

Substance	Approximate concentration in g per 100 cm^3			
	Colostrum	Human milk	Cow's milk	Blood plasma
Protein	8.0	2.0	4.0	7.0
Lipid	2.5	4.0	4.0	0.5
Milk sugar	3.5	8.0	5.0	(glucose) 0.1

antibodies, which help the baby overcome many different infections (this is more important in poor countries where infectious diseases such as cholera and typhoid are common)

○ it is convenient because no bottles, teats and other equipment are needed
○ it speeds the return of the uterus to its pre-pregnancy size
○ it creates a close bond between mother and baby which is important for the emotional well-being of the baby
○ both mother and baby usually enjoy it

Breast-feeding may be impossible because the mother does not produce enough milk or cannot be with her baby all the time (a bottle-fed baby can be fed by someone other than its mother). If a mother is ill, overworked or anxious, bottle-feeding removes a possible source of stress. Another advantage of bottle-feeding is that everyone knows how much milk the baby has had.

Many babies in rich countries are bottle-fed on carefully prepared milk formulae and thrive on it. In these countries families have all the equipment to keep bottles and teats sterile, and the water with which the formula is made up is free of infection. Unfortunately mothers in some poor countries have been encouraged to bottle-feed their babies even though they have plenty of breast milk, are with their babies all the time, do not have pure water or the means of keeping the formula sterile, and cannot afford to buy the formula in the first place.

24.5 Growth

At birth there are drastic changes in the way a baby gets its oxygen and its food: both are vital for its growth. For the first few days after birth the baby loses weight. After that it gains weight rapidly, particularly in the first two months. Figure 24.2 shows the rate of growth of a baby, weighed every four weeks, during its first year. Note that the graph plots *rate* of growth: the mass of different babies at birth and of different babies of the same age varies a lot; what is important is

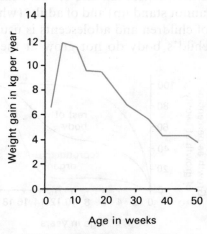

Figure 24.2   Weight gain from birth to 50 weeks

297

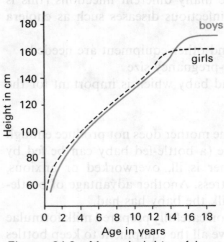

Figure 24.3 Mean heights of boys and girls aged 0 to 19

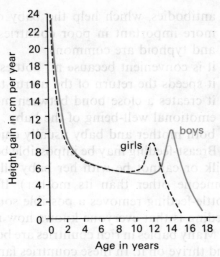

Figure 24.4 Height gains in boys and girls from birth to adulthood

that they should keep growing. The steady fall in the growth rate after 8 weeks means, not that the baby has stopped growing, but that it is growing less and less quickly. Figure 24.3, which plots the height of children from birth to adulthood, shows that the first year is the time of most rapid growth.

The height curves in Figure 24.3 show an interesting different between girls and boys. At thirteen the mean (average) height of girls is 5 cm greater than that of boys but by fifteen boys have overtaken girls and by eighteen boys are on average 11 cm taller than girls. The differences are shown clearly when the rate of growth is plotted against age as in Figure 24.4. Here it can be seen that girls start a growth spurt when they are about ten and boys start when they are about twelve. These growth spurts are related to puberty (see Section 24.6).

Neither mass nor height is a perfect measure of growth. Mass increases may be due to an increase, not in body tissue, but in stored lipid: in other words, people may be not growing but getting fatter. Height increases hide the fact that some people are tall and thin, others tall and broad. Mass is used to measure the growth of babies (who cannot stand up) and of adults (who have stopped growing in height). The growth of children and adolescents is usually measured in height.

Different parts of a child's body do not grow at the same rate. Figure 24.5

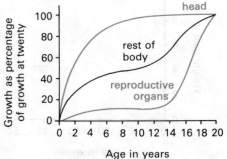

298   Figure 24.5   Growth of different parts of the body

shows the growth of the head, the reproductive organs and the rest of the body as percentages of their growth by the age of twenty. You can see that the head (and brain) grow fastest and almost reach adult size by the age of twelve. The reproductive organs grow slowly for the first twelve years, then quickly at puberty. The growth rate of the rest of the body more or less follows the growth rate in height (see Figure 24.3), which is not surprising because, apart from the head and reproductive organs, they are measuring the same thing.

24.6 Puberty

Humans have a long childhood before they become sexually mature and are able to reproduce. The growth of the reproductive system is delayed (see Figure 24.5) while, during their long childhood, they grow and learn. The start of sexual maturity, or the ability to reproduce, is called **puberty**. A girl is assumed to have reached puberty when she has her first menstrual period (though girls may have slight periods before they produce egg cells and menstruate properly). A boy is assumed to have reached puberty when he has his first ejaculation (though a boy may be able to produce sperm yet not have an ejaculation). The age of puberty varies greatly but is usually in the early teens or just before then. The changes occurring at about the time of puberty are known as the **secondary sexual characteristics**. The primary sexual characteristics are those we are born with: for example, ovaries in girls, testes in boys.

(a) Girls
The growth spurt occurs before puberty in girls. Hips get wider to make room for the developing fetus in pregnancy. (Girls are born with a large pelvic opening through which babies can pass.) Breasts get larger, the uterus and vagina get larger, and more lipid is stored under the skin making the body shape much rounder. Hair grows in the pubic region and the armpits. The lining of the uterus thickens, is shed and is replaced in menstrual cycles which at first may be irregular and do not always involve ovulation (see Section 22.3). Immature egg cells are present in the two ovaries of a girl long before her own birth; as soon as she is menstruating properly, one (at least) of these egg cells matures in each menstrual cycle. The sex hormone **oestrogen** controls the events of puberty in a girl.

(b) Boys
Puberty occurs on average only a few months later in boys than in girls but the growth spurt is about two years later. The testes, scrotal sac and penis enlarge. The voice box and vocal cords change to give a deeper voice. The skin becomes more greasy and sweaty. Muscles get larger and stronger. Hair grows in the pubic region, the armpits, on the face and chest, and elsewhere too. The testes manufacture sperms; glands in the reproductive system add the fluid that forms **semen**. The sex hormone **testosterone** controls the events of puberty in a boy.

Figure 24.6 shows how in both boys and girls the hormone secretion that controls sexual reproduction gradually increases throughout childhood. After puberty the sex hormones work differently in men and women. Testosterone

299

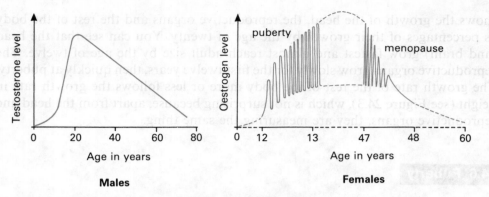

Figure 24.6   Testosterone and oestrogen secretion

secretion peaks at about the age of twenty and declines steadily until old age: it is unusual for men to have children after they are eighty. Oestrogen secretion rises and falls in each menstrual cycle, peaking in the late twenties and declining below puberty levels from the late forties onwards. This is the time, known as the **menopause**, when menstrual cycles and menstruation stop, after which women cannot have children.

24.7 Ageing

Cells taken from a young body and grown in a culture solution can divide as many as eighty times. Cells taken from older bodies can divide fewer times: the older the body, the fewer the cell divisions. Cancer cells are different: they go on dividing indefinitely. Cells are programmed to get old, not cancerous. Something has to go wrong for them to become cancerous.

As men and women get older

○ the lenses in the eyes get stiffer, it is more difficult to bring near objects into focus and people become long-sighted

○ the intervertebral discs flatten and people get shorter

○ calcium is withdrawn from bones, which become brittle and break easily (this happens earlier in women as oestrogen levels fall)

○ hair goes grey or white, pubic and armpit hair gets less (in men hair may disappear from the front and the crown of the head)

○ higher-pitch hearing is lost

○ the skin becomes less elastic

Many body processes decline after the age of 30. Figure 24.7 shows three reasons why older people's bodies are provided with less heat. As the amount of air breathed in is reduced, people get less oxygen for use in respiration. As the blood output from the heart is reduced, less oxygen and heat are transported in blood. As the resting rate of respiration is reduced, less heat is released in respiration. This loss of heat and a reduction in the efficiency of the temperature-control system can lead to hypothermia, particularly in old age when people do not move about so much and the muscles do not provide so much heat for the blood (see Section 20.3).

300

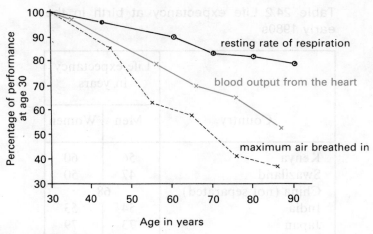

Figure 24.7 Bodily decline

Old people are often not as alert as they were when they were younger and may not realise how cold they have become. They need to keep rooms at a higher temperature and spend more money on heating than they used to, and this at a time when their incomes have probably fallen. Once their temperature falls below 35.0°C they will need an outside source of heat to get it back to normal again.

Studies have shown that taking regular exercise (see Sections 17.2, 18.4 and 19.4), keeping to a suitable diet (see Section 16.4), not smoking (see Question 19.4) and keeping the mind active (for example by travel or puzzles) can slow the ageing processes.

One of the most distressing diseases of ageing is **senile dementia (Alzheimer's disease)**, which affects one person in five over eighty but is very rare in people under sixty. Little was understood of this disease until the 1980s. As Figure 24.8 shows, it causes parts of the brain to waste away. The result is that the brain contains fewer nerve cells. The disease is thought to be caused by a faulty gene (see Section 25.3) which produces a damaging protein that gives rise to tangles of threads in the remaining nerve cells. Now that the cause of the disease is better understood, drugs are being developed to help patients suffering from it.

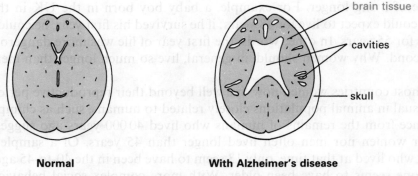

Figure 24.8 Normal brain and one with Alzheimer's disease (diagrams from brain scans)

301

Table 24.2 Life expectancy at birth in the early 1980s

Country	Life expectancy in years	
	Men	Women
Kenya	56	60
Swaziland	47	50
China (not separated)	68	
India	54	53
Japan	73	79
West Germany	69	76
Sweden	73	79
UK	70	76
USA	70	78
Brazil	67	73
Peru	57	60
USSR	66	75

Source: United Nations, *World Statistics in Brief*

24.8 Life expectancy

The age to which a person can expect to live is his or her **life expectancy**. It is worked out using death statistics that apply at that moment (it does not predict future trends in life expectancy). Table 24.2 shows you that life expectancy at birth is different for men and women and for different countries. It changes every so often as death rates change (in most countries people are living longer).

Life expectancy is different at different ages because some ages are more dangerous than others: a person who has passed the dangerous years has a better change of living longer. For example, a baby boy born in the UK in the early 1900s could expect to live for 48 years; if he survived his first year, he could expect to live for 55 years. In other words, the first year of life was more dangerous than the second. Why women should, in general, live so much longer than men is not clear.

In most countries women now live well beyond their reproductive period. This is unusual in animal populations closely related to humans, such as chimpanzees. Evidence from the remains of humans who lived 40 000 years ago suggests that neither women nor men often lived longer than 45 years. Of a sample of 152 adults who lived at that time, only 13 seem to have been in the 40-to-45 age group and none seems to have been older. With more complex social behaviour and more specialised tools, humans grew up more slowly and survived longer.

302

Q 24.1. What would you look for in an older man if you were trying to tell his age?

Q 24.2. Look at Table 24.1. In what ways is cow's milk different from human milk? Suggest what could be done to cow's milk to make the protein and sugar contents more like those of human milk.

Q 24.3. Look at Figure 24.7 and answer the following questions.
(a) Approximately what percentage of people's resting respiration rate at 30 is present at 80? How does the decline in the resting respiration rate relate to the increased danger of hypothermia in old age?
(b) Which of the three curves shows the greatest decline between 30 and 80? Why do you think this particular decline is so great?

GENETICS, CELL DIVISION AND GENETIC ENGINEERING

Genetics is the study of inheritance and of the similarities and differences between individuals caused by inheritance. A hundred years ago people had no idea how it came about that children looked like their parents and that brothers and sisters looked alike or how abilities (such as musical genius in the Bach and Mozart families) could be inherited. By the beginning of this century it was realised that there must be something in the egg cell and in the sperm which carried inherited material. The search began in earnest for what this inherited material could be.

Now we know that the inherited material is the DNA which is in the nucleus of every cell in the body and contains the coded instructions for making enzymes and for controlling all the activities of the cell. We also know how the DNA is passed on from fathers and mothers to their children (see Sections 14.1 and 22.2).

25.2 The nucleus and mitosis

Growth and the repair of tissue take place by the division of cells: one cell becomes two cells. The original zygote formed by the fusion of a sperm and an egg cell is a single cell. It divides into two. The two cells divide into four, the four into eight and so on. The process by which a nucleus normally divides is called **mitosis**. Cell division follows: each new cell has a nucleus exactly like that of the cell it was formed from. The nucleus of a cell contains the DNA, the inherited material. The DNA in the zytote is repeated in every cell of the body as the zygote becomes an embryo, the embryo becomes a fetus, the fetus becomes a baby, the baby becomes a child and the child becomes an adult. Finally half the DNA in the cell of each adult is passed on in a sperm or an egg cell to the next generation.

Figure 25.1, a photograph of a dividing onion cell, shows a nucleus in the middle of division by mitosis: at each end of the cell are a bunch of dark strands (dark because they have been stained to show them up). There are sixteen of   305

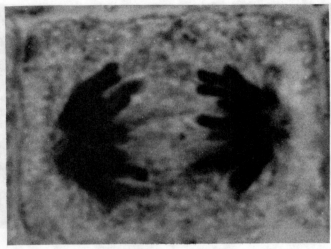

Figure 25.1   A dividing nucleus: a stage in mitosis

them at each end of this cell, though you cannot see all of them. They are called **chromosomes** and they contain and pass on inherited DNA.

A new nucleus will form around each group of chromosomes and a new cell will form around each nucleus. A few hours, days or weeks later each of the two new cells and their nuclei will themselves be ready to divide: in each nucleus there will still be only sixteen chromosomes but each chromosome will now be made up of two identical strands lying side by side.

During the process of mitosis each chromosome will separate into its identical strands and one of each of the identical strands will go to opposite ends of the cell. In the onion cell there will then be 32 single-strand chromosomes, sixteen single-strand chomosomes at each end of the cell. Each batch of sixteen is ready to form the nucleus of a new cell. This is a the stage shown in Figure 25.1.

Every time a body cell divides, its chromosomes split into two identical strands, one going into each new cell. Between each cell division and the next, the single-strand chromosomes become double-strand chromosomes: one strand of every double-strand chromosome goes into each of the two new cells at the next division, and so on. Of the sixteen chromosomes in an onion cell, eight came originally from the pollen and eight from the egg cell in the ovule. The two sets of eight can be recognised as eight pairs. The chromosomes of a pair look like each other and are different from all the other chromosomes.

In humans there are 46 chromosomes, two sets of 23, in each of our body cells. Of these, one set of 23 has come in the egg cell from the mother and the other set of 23 has come in the sperm from the father. It is now known that the chromosomes lie in special positions in each nucleus: at first it was thought that they lay randomly in the nucleus.

Every time a body cell divides, the 46 double-strand chromosomes split into 92 single-strand chromosomes: one of each pair of single-strand chromosomes goes to one new cell, the other strand to the other new cell. Between each cell division and the next, each of the 46 single-strand chromosomes becomes a

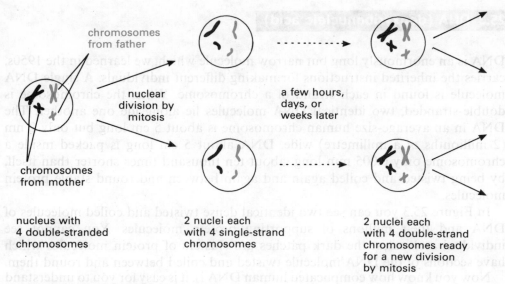

chromosomes
from father

nuclear
division by
mitosis

a few hours,
days, or
weeks later

chromosomes
from mother

nucleus with
4 double-stranded
chromosomes

2 nuclei each
with 4 single-strand
chromosomes

2 nuclei each
with 4 double-strand
chromosomes ready
for a new division
by mitosis

Figure 25.2   Nuclear division by mitosis

double-strand chromosome. Figure 25.2 shows mitosis in just four, two pairs, of the 46 human chromosomes.

Figure 25.3 is an electron **micrograph** (short for *microscope photograph*) of a double-strand chromosome (of a Chinese hampster) just before the strands separate. It can be seen that the two strands are still joined together. One strand will go to each of the two new cells. When the chromosomes are ready to divide again, each strand will have doubled itself and will look again exactly as it does in Figure 25.3.

Figure 25.3   Chromosome during division by mitosis

307

DNA is an enormously long but narrow molecule which, we learned in the 1950s, carries the inherited instructions for making different individuals. A single DNA molecule is found in each strand of a chromosome. When the chromosome is double-stranded, two identical DNA molecules lie alongside one another. The DNA in an average-size human chromosome is about 5 cm long but only 2 nm (2 millionths of a millimetre) wide. DNA about 5 cm long is packed inside a chromosome only 0.005 mm long, about ten thousand times shorter than itself, by being twisted and coiled again and again between and round stable protein molecules.

In Figure 25.3 you can see two identical dense twisted and coiled molecules of DNA and their millions of supporting protein molecules. You cannot see individual molecules: the dark patches are clusters of protein molecules which have sections of the DNA molecule twisted and coiled between and round them.

Now you know how complicated human DNA is, it is easy for you to understand how everyone (except identical twins, triplets, etc.) can have different DNA. What is called **genetic fingerprinting** identifies people from their DNA. It has nothing to do with fingerprinting. It is called genetic fingerprinting because it identifies people as reliably as real fingerprinting does. Not only can genetic fingerprinting (like real fingerprinting) distinguish between any two or more people (except identical twins, triplets, etc.), but it can do so from any of the cells of the body: for example, from the cells in a piece of skin, a hair with its root, a sample of blood or a sample of semen. Such evidence is often found at the scene of a crime when fingerprints are not.

Because genetic fingerprinting is accepted as proof of identity in a court of law (unless the person concerned is an identical twin or triplet, etc.), it is used to settle disputes over relationships. An immigrant's right to remain in a country may depend on proof that he or she is the child of parents who already have citizenship. A woman may be able to claim support for her child from its father only if she can prove he is its father. Occasionally babies are mixed up in hospital before the parents have got to know them and only science can decide which baby belongs to which parents. In the past blood tests have been used to decide such disputes but they could not always say with certainty what the relationships were. Genetic fingerprinting will be able to say with certainty what relationships are provided only that there are no identical twins involved.

(a) Bacterial DNA

The DNA in a bacterial cell is different from that in a human cell:

○ there is only one chromosome
○ the chromosome ends join up to form a long loop, not a strand
○ the DNA is twisted and coiled in the chromosome but is not wound round protein molecules
○ there is no nucleus and the chromosome lies in the cytoplasm

The bacterial DNA molecule is only about 1 mm long, but that is about a thousand times longer than the bacterial cell itself: only repeated twisting and coiling get it all in.

Some bacterial cells contain, in addition to the long chromosome loop, a small circular piece of DNA called a **plasmid**. The DNA in this plasmid can be opened up and pieces of DNA belonging to quite different organisms, including humans, can be spliced into it before it is closed up again. This process makes **recombinant DNA** and is part of a larger process called **genetic engineering** (see Section 25.4).

Figure 25.4 shows coiled and twisted DNA spilling from a damaged bacterium. A circular plasmid is about the size of the small loop shown by the arrow.

(b) Genes

The term **gene** was first used in 1909 to describe whatever it was that controlled an inherited feature: no one knew then what a gene was made of. Within five years scientists realised that genes were carried in chromosomes. Without having seen genes and without knowing what they were made of, they started to make the first chromosome maps of fruit flies showing which genes were grouped together on a chromosome and how close to one another or how far apart from one another they were. They worked all this out by studying inherited characteristics such as wing shape and body colour. In the 1940s scientists discovered that a gene controlled inherited characteristics by making a single protein, usually an enzyme.

After the discovery of DNA in the 1950s, scientists realised that a gene was a short piece of the long DNA molecule. Scientists are beginning to learn the positions of individual genes in the DNA. One day they will be able to cut out

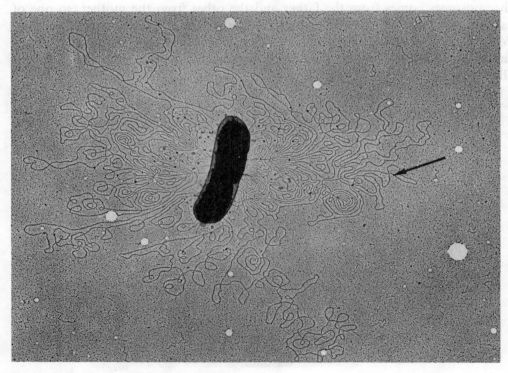

Figure 25.4   Bacterial DNA (× 5000)

faulty genes and replace them with perfect ones. At present there are only a few genes that they know anything about. Although a virus is known which contains only three genes in its DNA, bacterial DNA contains thousands and human DNA millions of genes.

Examples of single human genes that we know something about are those that control

- ○ the ABO blood-group system
- ○ phenylketonuria (PKU)
- ○ cystic fibrosis   ⎫ inherited diseases
- ○ sickle-cell anaemia   ⎭

Being able to roll one's tongue (as shown in Figure 25.5) is usually inherited as though it is controlled by a single gene. It is sometimes inherited in a slightly more complex way. Most inherited characteristics are controlled by more than one gene.

(c) Alleles

A gene can have two or more forms, called **alleles**. The differences between alleles are in the chemical structure of their DNA. As a result different alleles of a gene produce slightly different enzymes. Genes are arranged lengthwise along the DNA molecule in a chromosome. The two alleles of a gene are in exactly the same position in the two chromosomes of a pair; the two alleles may be either the same or different.

Remember that, of the 46 chromosomes in every human body-cell nucleus, 23 have come from the father and 23 from the mother. Just as one of each pair of chromosomes comes from the father and the other from the mother, so one of each pair of alleles of a gene comes from the father and the other from the mother. Whether the two alleles are the same or different, they control the same inherited characteristic. There follow some examples of human genes and their different alleles.

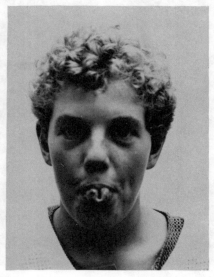

310   Figure 25.5   Tongue rolling

(i) Tongue-rolling

The effects of the tongue-rolling gene and its alleles we can all see for ourselves: some people can roll their tongues while others cannot. There are (effectively) two alleles of this gene, which we can write in shorthand as **T** (tongue-rolling) and **t** (non-tongue-rolling). The father will have given a chromosome containing DNA with one allele for the tongue-rolling gene; the mother will also have given a chromosome containing DNA with one allele for the tongue-rolling gene. One of two possible alleles, **T** and **t**, will be given by each parent to an individual.

> If both parents give a **T** allele, the individual has the two **TT** alleles and is a tongue-roller
>
> If both parents give a **t** allele, the individual has the two **tt** alleles and is a non-tongue-roller

What if one parent gives a **T** allele and the other gives a **t** allele? The individual has two different alleles, **T** and **t**, and might be expected to be half a tongue-roller. In fact the individual is a tongue-roller, just as much a tongue-roller as another individual with two **T** alleles. The explanation is that the tongue-rolling allele (**T**) completely masks the non-tongue-rolling allele (**t**).

It is because the tongue-rolling allele is **dominant** that it is written with a capital **T**. The non-tongue-rolling allele, written with a small **t**, is said to be **recessive**. This means that it will not have an influence on the individual unless it is paired with another **t** allele.

All the possibilities of tongue-rolling inheritance are shown in Figure 25.6. The x sign is used to show a mating or cross between two individuals. When different alleles of a gene are present, it is usual to show the allele that masks the other first: for example, **Tt**. But the alleles may be written in either order: **Tt** and **tT** both show a tongue-rolling individual.

When there are two different alleles of the same gene, one may be dominant and the other recessive. When there are two different alleles of the same gene and neither is dominant nor recessive, they are said to be **codominant**. If you cross red and white antirrhinum flowers and get an antirrhinum with pink flowers, you have an example of codominance. The ABO blood-group system and sickle-cell anaemia are other examples of codominance (see below).

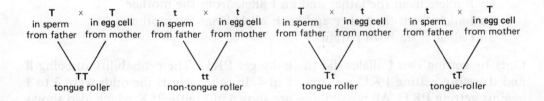

Let T = tongue-rolling allele
and t = non-tongue-rolling allele

Figure 25.6 Tongue-rolling inheritance

311

When the two alleles of a gene are the same, for example **TT**, or **tt**, the individual is **homozygous** for that gene; when two alleles of a gene are different, for example **Tt**, the individual is **heterozygous** for that gene.

The alleles in an individual for a particular gene are called the **genotype**: **TT**, **tt** and **Tt** (or **tT**) are three different genotypes for the tongue-rolling gene. The actual characteristics of an individual are called the **phenotype**. For the three genotypes **TT**, **tt** and **Tt** (or **tT**) there are only two phenotypes: tongue-rollers and non-tongue-rollers.

(ii) Phenylketonuria (PKU)

PKU is an inherited disease (see Section 24.3) in which the enzyme that can break down a particular amino acid (phenylalanine) is not working properly. If PKU is not discovered at birth or soon afterwards, it will cause brain damage when that amino acid gets concentrated in the body. It is more common among white people.

In shorthand the allele for making the enzyme that prevents PKU is **F** and the allele for making the faulty enzyme is **f**. Because lack of the enzyme causes PKU, it would be natural to use **P** and **p**. But a capital **P** and small **p** when written by hand are difficult to tell apart: it is better to choose letters that look quite different as capitals and small letters: for example, **A** and **a** and **B** and **b** are suitable, but **C** and **c**, **K** and **k** and **U** and **u** are not. **F** and **f** were chosen for PKU because the Ph with which its full name begins is pronounced like an F. But any pair of letters could have been chosen provided that the capital and small letters looked different.

We do not need to write the names of the gametes (sperm and egg cell) each time, and instead of 'father' and 'mother', we can use the symbols ♂ for male and ♀ for female.

A baby can have PKU even though both its parents are healthy. To understand how this happens, look at Figure 25.7.

The baby is **ff** and both its healthy parents are **Ff**. Both its parents have the recessive **f** allele causing PKU, but in both of them it is masked by the dominant **F** allele for preventing it. Both its parents are **carriers** of the disease without having it.

The baby was unlucky to get an **f** allele from both its **Ff** parents. The baby could have had

> an **F** allele from both parents
> an **F** allele from the father and an **f** allele from the mother
> an **f** allele from the father and an **F** allele from the mother
> an **f** allele from both parents

Only by getting two **f** alleles did the baby get PKU. The probability of being **ff** and therefore getting PKU was only 1 in 4. In other words the odds were 3 to 1 against getting PKU. All possibilities are shown in Figure 25.8, which also shows how to draw a genetic diagram in which all possible combinations of alleles from the two parents are shown.

Figure 25.8 also shows that you should draw a circle around the alleles in a gamete. They cannot then be confused with alleles of the parents or offspring.

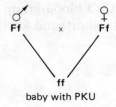

Let **F** = normal allele preventing PKU
and **f** = allele causing PKU (phenylketonuria)

Figure 25.7   PKU inheritance by a baby of carrier parents

Look at the genotypes of the offspring in the diamond checkerboard. Can you see that there are three genotypes for this gene? It may look like four, but **Ff** and **fF** are two ways of writing the same thing. Can you see that there are two phenotypes for this gene? They are a phenotype without PKU (**FF** and **Ff** and **fF**) and a phenotype with PKU (**ff**).

The symbols and layout used in Figure 25.8 should be used for all animal and plant matings and crosses. A complete layout of a genetic cross shows

> Parental phenotypes
> Parental genotypes
> Gametes
> Offspring genotypes
> Offspring phenotypes

in that order.

(iii) The ABO blood-group system

When there are three or more alleles (**multiple alleles**) of a gene, the gene itself is given a capital letter. The alleles are then shown by capital letters if they are dominant or codominant, and by small letters if they are recessive, at the top right of the gene's capital letter.

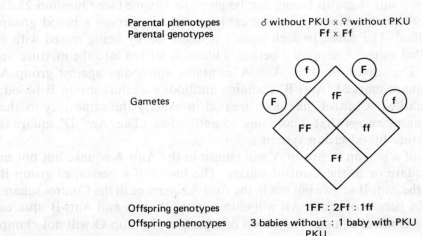

Parental phenotypes	♂ without PKU x ♀ without PKU
Parental genotypes	Ff x Ff
Gametes	
Offspring genotypes	1FF : 2Ff : 1ff
Offspring phenotypes	3 babies without : 1 baby with PKU PKU

Figure 25.8   PKU inheritance

313

The gene that controls the ABO blood-group system (**I**) has three alleles (an example of multiple alleles). The shorthand for these alleles is

I^A for the A blood group
I^B for the B blood group
I^o for the O blood group

The I^A and I^B alleles are codominant: both are dominant to the recessive I^o allele. Though three different alleles exist, there can be only two in any one genotype. With three alleles there are six possible genotypes (six combinations of two):

$I^A I^A$
$I^A I^o$
$I^A I^B$
$I^B I^B$
$I^B I^o$
$I^o I^o$

All these genotypes exist: $I^A I^A$ and $I^A I^o$ make the person blood-group A (because the I^o allele is recessive); $I^B I^B$ and $I^B I^o$ make the person blood-group B; $I^A I^B$ makes the person blood-group AB; $I^o I^o$ makes the person blood-group O. The blood-group AB is both blood-group A and blood-group B. This is because the alleles I^A and I^B are codominant. You can see that the ABO blood-group system has four different phenotypes, A, B, AB and O. In other words, in the ABO blood-group system there are four different blood groups.

You must be one of the four blood groups, A, B, AB or O. Which you are is of no importance in everyday life. Your blood group becomes important only if you need a blood transfusion or if you give blood. A-group blood must never be given to a person without A-group blood; B-group blood must never be given to a person without B-group blood. A-group blood can be given to an A-group or an AB-group person but not to a B-group or an O-group person. AB-group blood can be given only to an AB-group person. O-group blood, because it has neither A-group nor B-group blood, can be given to anyone (see Question 25.2).

Figure 25.9 shows a blood-grouping card by which a person's blood group can be identified. The patch in each square is dissolved by being mixed with a drop of distilled water. A drop of a person's blood is stirred into the mixture on each square. The square marked 'Anti-A' contains antibodies against group A blood; the square marked 'Anti-B' contains antibodies against group B blood; the square marked 'Control' has been treated in exactly the same way as the other two squares *except* that it contains no antibodies. (The 'Anti-D' square is a test for another blood-group system.)

The blood of a person of group A will clump in the Anti-A square but not in the Anti-B square or in the Control square. The blood of a person of group B will clump in the Anti-B square but not in the Anti-A square or in the Control square. The blood of a person of group AB will clump in the Anti-A and Anti-B squares but not in the Control square. The blood of a person of group O will not clump in any square.

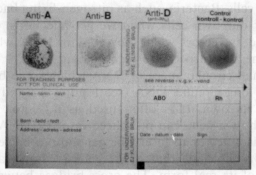

Figure 25.9   Blood-grouping card

The blood-grouping card in Figure 25.9 has been used. In the Anti-A square, the blood has formed many small clumps (dark patches); so has the blood in the Anti-B square, though there are fewer of them and they do not show up so clearly. Although blood in the Control square is denser in parts than others, it has not clumped. (Blood in the Anti-D square has also clumped.) Because this blood has reacted with A and B antibodies, it must contain the A and B substances. The person is blood-group AB.

25.4 Genetic engineering

It is now possible to separate the short piece of DNA which is a single gene from the long molecule of DNA in which it lies. The gene can then be moved into another molecule of DNA, which can be put into another organism. When the technique is successful, the moved gene will carry out what it is coded for in the new organism. The technique is called **genetic engineering**.

Even under today's powerful microscopes, genes cannot be seen well enough to be recognised. Genetic engineers start with strands of DNA which they know contain genes including the one they want. They cut the strands of DNA into pieces with the use of a special (restriction) enzyme. They then use a variety of chemical and physical processes to sort out the pieces until they find pieces of DNA each containing only the gene they want. The **vector DNA**, the DNA that will be used to move the gene into the organism, is cut open with the same enzyme. Because the same enzyme is used, it leaves similar **sticky** cut ends on the DNA of both the gene DNA and the vector DNA. Genes and vector DNA are mixed together with another enzyme (a ligase) which joins sticky ends together.

It is a hit-or-miss process. Not all the sticky ends will join together in the ways the genetic engineers want, but some will. Some of the genes they want will be **spliced** into the vector DNA to form what is called **recombinant DNA**. Recombinant DNA can be inserted into the cells of an organism and the organism can be cloned (reproduced many times).

What can be used as vector DNA? Bacteria and viruses both have the ability   315 to enter into other cells. Bacteria, such as nitrogen-fixing bacteria, can enter plant cells; viruses enter our own cells when we get infections. Certain bacteria and viruses can be used as vectors when they contain DNA which has been spliced with a selected gene and turned into recombinant DNA.

A plasmid, a small circular piece of DNA which some bacterial cells contain in addition to the long circular DNA (see Figures 2.3 and 25.4), can enter into bacterial cells on its own. It is often used as a vector. Figure 25.10 shows diagrammatically how a gene can be spliced into the DNA of a plasmid, which can then be used as a vector to insert the gene into a bacterium.

Part of the DNA molecule that is the gene is shown as a ladder: this is a simplified version of its real form. The cross-bridges between the two sides are linked by hydrogen bonds. Similarly the cross-bridges on the plasmid are linked by hydrogen bonds. These bonds are weaknesses in the DNA molecule when it splits and joins up again.

Genetic engineering is used to produce human insulin. Insulin, a hormone which removes sugar from the blood, has to be taken regularly, often daily, by some diabetics. In the past insulin had to be extracted from the pancreas (a gland near the small intestine) of animals. Because it did not have exactly the same composition as human insulin, its use could lead to medical complications in the patient.

The gene that codes for human insulin is extracted from human DNA and put into plasmid DNA (the vector). The recombinant plasmid DNA is then put into the harmless bacterium *Escherichia coli*. It does not matter that *E. coli* already has its own DNA. It now obeys the coded instructions both in its own DNA and in the recombinant plasmid DNA. It therefore makes human insulin. The next step is to clone *E. coli* containing the recombinant plasmid DNA. Vast numbers of these harmless bacteria are produced, all of which make human insulin. The production of human insulin has become a successful industrial process.

Genetic engineering has exciting possibilities. Can bacteria be programmed to

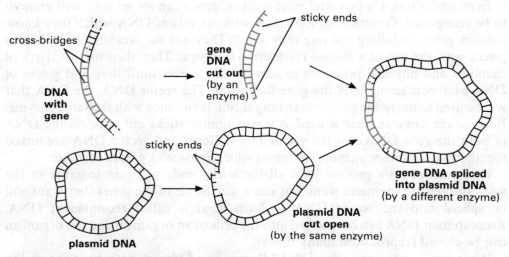

316   Figure 25.10   Formation of recombinant DNA

produce the antibodies that we need against diseases? Can bacteria be programmed to produce an AIDS vaccine? Can the bacteria that fix nitrogen gas from the atmosphere be put into cereal crops so that nitrogen fertilisers are no longer needed? Can we reduce the number of handicapped children by substituting normal alleles for harmful alleles in human embryos?

25.5 Family pedigrees

Race horses and dogs have **pedigrees** and so do people. A pedigree is a record of an individual's relatives, especially the ancestors from whom he or she has descended. Members of royal families, where relationships are important because of succession to thrones and titles, have well kept pedigrees going back many generations. Most of us can work back only a few generations. Pedigrees are of particular interest to geneticists, who are able to work out the inheritance of certain diseases from them and to help people by **genetic counselling** (see Section 26.6).

Symbols are used in pedigree analysis, just as they are in genetic diagrams (see Section 25.3), but the ♂ and ♀ signs are replaced by squares (for males) and circles (for females). Figure 25.11 gives the pedigree of two children both with **cystic fibrosis** and shows some of the symbols that are used.

Cystic fibrosis is an inherited disease in which secretions from glands, especially glands that secrete mucus in the breathing tubes and gut, are thick and sticky. The allele of the gene that causes cystic fibrosis is recessive to the allele of that gene that keeps the glands working properly. To get the disease a child must therefore inherit a cystic-fibrosis allele from both parents. If neither parent has cystic fibrosis yet a child has it, both parents must be carriers.

Individuals in a pedigree can be named or numbered so that they may be referred to. From the pedigree in Figure 25.11 a geneticist can work out what the probabilities are that other members of the family have a recessive cystic-fibrosis

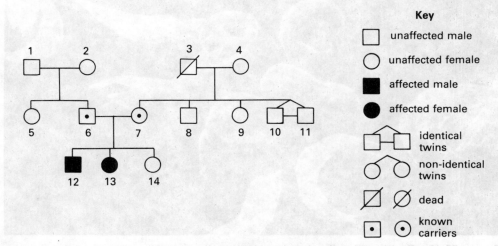

Figure 25.11 Family pedigree of two children with cystic fibrosis

allele. Look at Figure 25.11. Children 12 and 13, who have cystic fibrosis, must have inherited the recessive allele from both parents. Therefore both parents must be carriers. To be carriers they must both have inherited one cystic-fibrosis allele from one of their parents. Either individual 1 or 2 is a carrier or both are, and either individual 3 or 4 is a carrier or both are. At least two of the grandparents of children 12, 13 and 14 must be carriers; from this pedigree we do not know which.

Other members of this family now know that they too may be carriers of the cystic-fibrosis allele: the probability that each of the individuals 5, 8, 9, 10 and 11 is a carrier is at least 1 in 2 (1 to 1) because at least one of their parents had one healthy allele and one cystic-fibrosis allele. If both the parents of the individuals 5, 8, 9, 10 and 11 were carriers (and there is no way of telling if they were), the probability that the individual is a carrier is 2 in 3 (2 to 1).

Another inherited disease is **sickle-cell anaemia**. It gets its name from the fact that, in sufferers from the disease, many of the red blood cells have a peculiar crescent shape like a sickle (a tool for cutting grass). Figure 25.12 shows red blood cells from a person with sickle-cell anaemia. The sickle cells do not carry oxygen as well as normal red blood cells. Sufferers from sickle-cell anaemia often have pains in their joints (because the abnormal cells get stuck in small capillaries); they get coughs and colds more easily than other people; they may sometimes feel tired and ill. Sickle-cell anaemia is caused by a fault in the haemoglobin molecule. It is more common among black people.

Hb^A is the symbol used to show the allele for normal haemoglobin and Hb^S is the symbol for the allele for sickle-cell haemoglobin. The symbol Hb is treated as though it were a single capital letter. The two alleles Hb^A and Hb^S are codominant. (Capital letters with small capital letters top right are used for both multiple alleles and codominant alleles.) A person with two codominant alleles shows some of the characteristics of both (like the pink antirrhinum with red

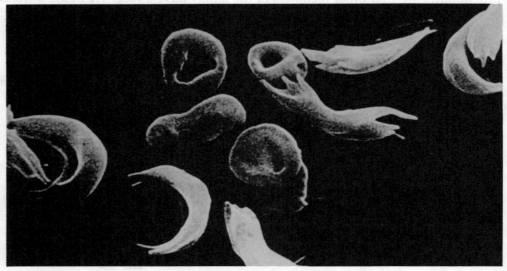

318   Figure 25.12   Red blood cells showing sickle-cell anaemia

and white alleles). Fortunately people with the genotype **Hb**^{A}**Hb**^{S} show only slight characteristics of the **Hb**^{S} allele. If their blood is tested, some of their red blood cells can be made to sickle. But they lead healthy lives. They are known as carriers because for practical purposes it is as though the **Hb**^{S} allele is recessive to the **Hb**^{A} allele. Only people with the genotype **Hb**^{S}**Hb**^{S} suffer from sickle-cell anaemia.

Questions

Q 25.1. Use the symbols for the blood-group alleles shown in Section 25.3.
(a) Give the possible genotypes of someone with blood-group B.
(b) A man and a woman, who are both blood-group A, have a child who is blood-group O. Explain how this is possible.
(c) A woman who was blood-group O was certain that a man who was blood-group B was the father of her child, who was blood-group O. The man said this was impossible. Was he right? Explain your answer.

Q 25.2. Copy Table 25.1 and complete it to show by ticks what blood can be given to the patients listed on the left. The first column has been done for you and shows that blood of group A can be given only to patients who are blood-groups A or AB.

Table 25.1

	Blood group of donor (person giving blood)			
	A	B	AB	O
Blood group of patient A (person receiving blood) A	✓			
B				
AB	✓			
O				

Q 25.3. Look at Figure 25.11 and answer the following questions.
(a) What is the sex of the identical twins?
(b) How many grandparents of children 12, 13 and 14 are still living?
(c) How was it worked out that both parents 6 and 7 are carriers of cystic fibrosis?
(d) What is the probability that child 14 is a carrier of cystic fibrosis?
(e) Could any of the people who are shown as 'unaffected' be carriers of cystic fibrosis? Explain your answer.

Q 25.4. The affected people in the pedigree shown in Figure 25.13 have a *dominant* allele which causes a form of dwarfism (they grow into dwarfs).
(a) How do you know that all the unaffected people are homozygous for the normal allele that prevents this form of dwarfism?

319

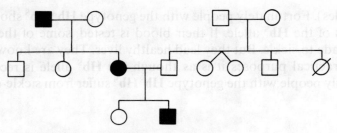

Figure 25.13 Family pedigree for Question 25.4

(b) Use the key in Figure 25.11 to help you answer the following questions.
  (i) How do you know that the affected grandfather is heterozygous for this form of dwarfism?
  (ii) Let **D** = the allele for this dwarfism and **d** = the normal allele. What is the genotype of the mother of the two grandchildren?
  (iii) What do you know about the aunts and uncles of these two children?

INHERITANCE AND VARIATION

26.1 Introduction

It is through the genes in our chromosomes that we inherit characteristics from our parents. It is through one of the two alleles of each gene in each chromosome that we pass on to our children in gametes half of what we have inherited.

26.2 Gamete formation

Sperms are formed in the testes, egg cells in the ovaries. Before an egg cell and a sperm fuse to form a zygote, the chromosome number, which is 46 in our body cells, is halved to 23 in each gamete. The process by which this is done is called **meiosis**. The 23 chromosomes in a gamete are all different. They form a **haploid** (single) set of chromosomes; when there are 46 chromosomes, as there are in all the body cells, they form a **diploid** (double) set. In number the haploid set is always half the diploid set. Different organisms have different numbers of chromosomes in their sets. For example, the onion cell that was dividing in Figure 25.1 had sixteen chromosomes in its diploid set: onion gametes (in the pollen tube and ovule) have haploid sets of eight.

Figure 26.1 shows the 46 chromosomes in the nucleus of a human male: the chromosomes were photographed during a division by mitosis (as shown in the bottom left of the picture): the photographs of the individual chromosomes were cut out and arranged according to size (largest top left, smallest bottom right). One pair is not of equal size: its chromosomes are called X and Y: they are put in their correct positions according to size. A human female has two X chromosomes instead of an X and a Y (see Section 26.3). This kind of picture of chromosomes is called a **karyotype**.

At meiosis, when the gametes are about to be formed, each diploid set of 46 chromosomes divides into two haploid sets of 23. In a male, four sperms are formed from each diploid cell: each has a haploid set of 23 chromosomes. Each chromosome of a pair, including the X and Y chromosomes, ends up in a different sperm.

In a female, there is usually only one egg cell produced by each diploid cell: one haploid set of 23 chromosomes, including an X chromosomes, ends up in the egg cell; the other sets of haploid chromosomes get lost.

At the start of meiosis, the pairs of chromosomes come to lie together and twist around each other. Each chromosome is a double strand, as it is at the beginning 321

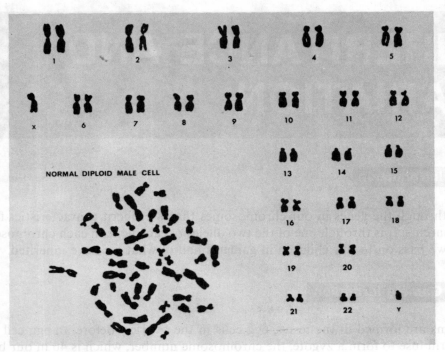

Figure 26.1 Human karyotype

of mitosis, and the strands become so closely interwoven that they exchange DNA. Then, as meiosis proceeds, the chromosomes in each pair repel their partners and go to opposite ends of the dividing cell. The result is 23 double-strand chromosomes, a haploid set, at each end of the cell. In a male, one set contains an X chromosome and the other a Y chromosome; in a female each set contains an X chromosome.

Notice the difference between what has so far happened in meiosis and what happens in mitosis (see Section 25.2). In mitosis the 46 double-strand chromosomes split into 92 single-strand chromosomes: 46 single-strand chromosomes go to each end of the cell ready to form two new cells each with 46 single-strand chromosomes (which will develop into double-strand chromosomes). So far in meiosis the 46 double-strand chromosomes have divided into two sets and gone to opposite ends of the cell. While in mitosis there is a diploid set of 46 single-strand chromosomes at the each end of the cell, at this stage of meiosis there is a haploid set of 23 double-strand chromosomes at each end of the cell.

You may be able to guess what happens at the next stage of meiosis. At each end of the cell a nucleus forms round a haploid set of 23 double-strand chromosomes and the cell divides into two. The number of chromosomes in a cell has now been halved. But each chromosome is still in the form of a double-strand. Another division of each nucleus and each cell takes place to reduce the double-strand chromosomes to single-strand chromosomes: this division is very much like mitosis. Figure 26.2 shows the main events of meiosis in the male, in whom the final cell divisions are complete. Again only four, two pairs, of the 46 human chromosomes are shown.

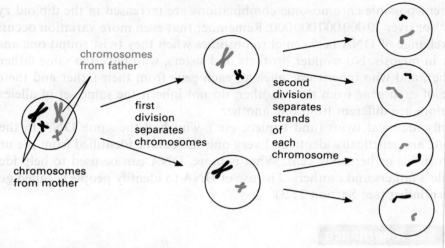

chromosomes from father

chromosomes from mother

first division separates chromosomes

second division separates strands of each chromosome

Figure 26.2   Nuclear division by meiosis

The cells formed after the final division of meiosis are ready to become gametes. In the male they all become sperms. In the female only one of the two cells that formed after the first division goes on to divide again; only one of the cells formed after the final division becomes an egg cell, and this final division takes place only if a sperm enters the egg cell.

Strictly speaking, what the sperm enters is not yet an egg cell. It is a cell with 23 double-strand chromosomes about to divide to form an egg cell. It is a cell like that shown after the first division in Figure 26.2. As soon as the sperm enters this cell, the second division takes place: the cell gets rid of a single haploid set of single-strand chromosomes. The set that is got rid of does not form another egg cell; it is lost. Look at Figure 14.6: the two cells at the surface of the egg cell contain the discarded chromosome sets of the first and second divisions of meiosis. What is left is a fertilised egg cell or zygote with 23 single-strand chromosomes from the egg cell and 23 from the sperm. Thus it has the normal number of 46 single-strand chromosomes (which will develop into double-strand chromosomes). Of every pair of chromosomes in the body cells, one has come from the father and one from the mother. When meiosis takes place, the father's chromosomes and the mother's chromosomes move randomly to either end of the cell. The number of different ways that the 23 chromosomes, one from each pair can come together at each end of the cell is 2^{23}, which is 2 multiplied by itself 22 times, which works out at 8 388 608. In other words, there are over eight million possible ways in which one chromosome from each of the 23 chromosomes pairs can combine to form the haploid set of 23 in any one gamete.

Even that is not a lot when compared with the 200–300 million sperms likely to be present in a single ejaculation. But the fact that there is exchange of DNA during meiosis means that all the sperms are likely to have different DNA.

Meiosis is therefore one of the ways in which **genetic variation** is brought about. Variation is further increased when two gametes, sperm and egg cell, each of which has over eight million possible chromosome combinations, fuse. The   323 different possible chromosome combinations are increased in the diploid zygote to 2^{46} or over 70 000 000 000 000. Remember that even more variation occurs due to exchange of DNA between chromosomes when they twist round one another early in meiosis. No wonder brothers and sisters, who have the same father and mother, and who inherit one allele of each gene from their father and the other allele of each gene from their mother, do not inherit the same set of alleles and therefore are different from one another.

Only identical twins (and triplets, etc.), who had the same DNA in the one zygote, are genetically identical. Every one else can be identified from the unique DNA in his or her body cells. What is more, DNA can be used to help identify people's fathers and mothers. The use of DNA to identify people is called **genetic fingerprinting** (see Section 25.3).

26.3 Sex inheritance

A person's sex is controlled by a few genes or possibly by only one gene on the Y chromosome. A person who has this gene or these genes is a male. Look at the karyotype in Figure 26.1. The X and Y chromosomes are the ones that control sex. The human male has one large X and one small Y chromosome which do not form a pair of identical shape. The human female has two large X chromosomes which do form a pair of identical shape. At meiosis in the male, an X chromosome goes to one nucleus and a Y chromosome goes to the other: when a sperm is formed, its haploid nucleus contains either an X chromosome + 22 other chromosomes or a Y chromosome + 22 other chromosomes. The two different kinds of sperms, one kind with an X and the other kind with a Y chromosome, should be present in equal numbers.

When meiosis takes place in the human female to form an egg cell, the paired X chromosomes behave just like the other paired chromosomes: one X chromosome goes to one end of the cell with another 22 chromosomes and the other X chromosome goes to the other end of the cell with another 22 chromosomes. More than a thousand billion sperms may be formed during the lifetime of a man, but only a few egg cells are formed in the lifetime of a woman.

In a man both the divisions of meiosis take place a few days before sperms are formed in the testes. In a woman the first division of meiosis starts in her ovaries before she is born, when she is still a fetus in her mother's uterus; when she is born, she already has a store of immature egg cells formed in her ovaries. Egg cells do not finish their first division of meiosis, begun before birth, until they mature (usually one at a time) in the early part of a menstrual cycle (see Section 22.3). The second division of meiosis, which reduces double-strand chromosomes to single-strand chromosomes, does not take place in a woman unless and until a sperm cell enters an egg cell.

All the egg cells produced by a woman contain an X chromosome; half the sperms produced by a man contain an X chromosome and half contain a Y chromosome. Whether an X-carrying sperm or a Y-carrying sperm fertilises the egg cell determines the sex of the zygote and of the child that develops from it. This is shown in Figure 26.3.

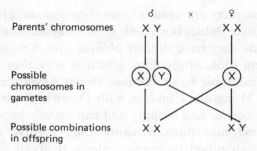

Parents' chromosomes ♂ × ♀
X Y   X X

Possible
chromosomes in
gametes
X Y   X

Possible combinations
in offspring
X X   X Y

Figure 26.3   Sex inheritance

Look at Figure 26.3. You can see that, if equal numbers of X sperms and Y sperms are produced, the numbers of girl and boy babies born should be equal. In fact they are not equal. In the UK 1054 boys are born for every 1000 girls. The reason for this is unclear, but there are several possibilities: more Y-carrying than X-carrying sperms may be made; more Y-carrying sperms may reach an egg cell; Y-carrying sperms may be better at fertilising egg cells; girl embryos may be more likely to be spontaneously aborted (see Section 23.6).

A hundred years ago, although more boys were born, they were more likely than girls to die in childhood and to die later in wars. Surviving men were also more likely to emigrate. Instead of more men in the UK, there were more women. Almost one-third of the women at that time did not marry, mainly because there were not enough men to go round. With improved medical knowledge, no major wars and little emigration, there are now over a quarter of a million more men than women in the 15–29 age group in the UK and the figure is rising. It is men who will find it harder to get married in future.

26.4 Mutations

It has been assumed so far that DNA never changes. This is not quite true. Occasionally there is a sudden change in the DNA called a **mutation** (which means a change).

(a) Chromosome mutations

A major mutation takes place when something goes wrong at cell division and too few or too many chromosomes go into the newly formed cells. This has a drastic effect if it happens in a gamete, a sperm or an egg cell, which is involved in reproduction. Most zygotes which develop into embryos with more than 46 or less than 46 chromosomes abort naturally (see Section 23.6) in the early weeks of pregnancy.

About 45 per cent of early spontaneously aborted embryos and fetuses have an extra chromosome and about 20 per cent have an X chromosome missing. Even so, about 500 babies a year are born in the UK with chromosome mutations: many have an abnormal number of sex chromosomes (for example XXY, XYY, X and XXX) but one of the commonest chromosome mutations has nothing to do with the sex chromosomes and causes **Down's syndrome**.

In Down's syndrome there are usually three chromosome 21s (see Figure 26.1) instead of the normal two. Babies born with Down's syndrome can be recognised soon after birth because they have slightly oblique eyes, a round head and a flat nose. This makes them look Mongolian, which is why they used to be called mongols: this is an unsuitable name because Down's syndrome has nothing to do with people from Mongolia. Children with Down's syndrome are mentally handicapped but are cheerful and friendly and can usually look after themselves.

All the chromosome mutations, because they involve extra or missing chromosomes, can be identified by amniocentesis at about five months into a pregnancy and by chorionic villus sampling even earlier (see Section 23.9). If chromosome mutations are found at these early stages in a pregnancy, parents in the UK are offered the chance of having a (legal) induced abortion and of beginning again with a new pregnancy.

(b) Gene mutations

While a chromosome mutation can be seen when dividing cells are looked at under an ordinary light microscope, a gene mutation involves such a small change in the DNA of a cell that it cannot be seen even with an electron microscope. The inherited diseases, such as PKU, cystic fibrosis and sickle-cell anaemia, are all believed to have arisen when a normal piece of DNA was slightly changed creating a faulty allele.

In 1986 scientists developed a test, using what is called a **gene probe**, which can determine as early as eight weeks into a pregnancy whether a fetus carries two recessive cystic-fibrosis alleles. If it does, the parents can again be offered an induced abortion before the pregnancy is well established. These tests are available to pregnant women who are known to be at risk of having a baby with cystic fibrosis. Gene probes can even identify recessive cystic-fibrosis alleles in people without the disease. This means that healthy people with cystic fibrosis in their families can find out if they are carriers before deciding whether to have children. Gene probes for many more inherited diseases are being developed.

(c) Mutations as a source of variation

Nearly all mutations are harmful to the organisms in which they occur, but occasionally there are beneficial mutations. When bacteria suddenly develop a resistance to an antibiotic, it is because a mutation has created a new allele which produces an enzyme which breaks down the antibiotic. This has happened many times among different bacteria with the result that many of the commonly used antibiotics, including penicillin, are becoming less effective. It is why antibiotics should not be used unnecessarily: the more they are used, the more the non-resistant bacteria are killed and the more the resistant bacteria multiply.

The DNA in some viruses consists of only a few genes: even these occasionally mutate. Vaccines and a person's own antibodies against flu and the common cold do not work for long because the viruses that cause flu and the common cold keep mutating. The viruses that cause polio are more stable. This is why you are offered inoculation against polio but not against flu and the common cold.

If it were not for mutations, there would be no great changes in organisms. The same inherited material would be shuffled endlessly at sexual reproduction and no new varieties of organisms would arise. It is mutations which give rise to completely new features, which may be valuable in crop and livestock selection (see Section 11.4). Without mutations there could not have been evolution of apes and humans from a common ancestor. Indeed there could not have been evolution from microscopic organisms to humans.

(d) Mutagens

Mutations in DNA can cause birth defects if they occur in gametes and cancer if they occur in body cells. Chemicals and radiations which are known to produce mutations in DNA are called **mutagens**. X-rays were once thought to be harmless but are now known to be mutagens. Much greater care is now taken by radiologists and technicians when handling X-rays and other radiations. X-rays are taken only when they are essential and the reproductive organs are always well shielded from radiation. In the nuclear industry, where radiations are produced, safety precautions are strict. Pollution by waste from the nuclear industry is dangerous because it can cause cancer (see Section 13.3).

26.5 Variation

This Unit and the previous one have been concerned with inherited variation of a particular kind: that caused by variation in genes and chromosomes. In the examples used a gene has had only two or, in the ABO blood-group system, three different alleles. Most variation is not as straightforward as this.

People are rarely the same height as their parents and their brothers and sisters. They are often not even between the heights of their parents. But tall parents do tend to have tall children. Height is therefore inherited but not through one gene with two alleles or even through one gene with three alleles. It is inherited through one gene with many alleles or, more likely, through several genes with many alleles. Height is also related to sex: the average man is taller than the average woman. Altogether height has a complex pattern of inheritance. Most inheritance is as complex as, or even more complex than, the inheritance of height.

Inheritance is not the only factor that controls our features. For example, adult height is also related to the kind and amount of food eaten during childhood. During the twentieth century the average height of men and women in the UK has risen steadily. Because their inherited characteristics cannot have changed so much in so short a time, this is evidence that some variation is due to influences during life (including life in the uterus). The increase in height must be due to better food, better living conditions and better medical care. If identical twins, who inherited exactly the same alleles controlling height, were brought up separately with different lifestyles, they could be very different heights: they would be the same genotypes but different phenotypes.

Height is but one of the many ways in which people vary. Look at Figure 26.4. There are people of different sexes, different ages, different heights, different builds and mass, different skin colours and different hair colours. Some are strong, some weak; some healthy, some unhealthy. All have different characters and abilities. Human variation is infinite.

Figure 26.4   Human variation

26.6 Genetic counselling

Down's syndrome, cystic fibrosis and sickle-cell anaemia are just three examples of many diseases which people are born with because they are unlucky enough to have the wrong alleles. Many more people, though themselves healthy, know that such a disease (or some inherited handicap other than a disease) runs in their family. They wonder if they themselves are carrying it and if their own children will inherit it. Nowadays, before deciding to have children, they can seek advice from a **genetic counseller**. A genetic counsellor is a doctor who tells future parents how great or small is the risk that their children will be born with an inherited disease and what they can do about it.

Suppose that a couple both have cystic fibrosis. This can easily happen because two people with cystic fibrosis are likely to go to the same place for treatment, to get to know each other and to fall in love. Cystic fibrosis is caused by a recessive allele of a single gene; therefore both the man and woman must be carrying two of these recessive alleles. Sadly it is certain that all their children will get two of these recessive alleles, one from each parent, and will suffer from cystic fibrosis.

On being told this, the couple may decide to have children nonetheless. Because they have the disease themselves, they understand it well and will feel well qualified to look after children who have it. On the other hand, because they themselves are handicapped, they may think it will be difficult for them to give similarly handicapped children all the help they need. Or they may not want to bring into the world children who will have all the difficulties that they themselves have had. They may therefore decide to have no children, to adopt children or to have AID (artificial insemination by donor) (see Section 22.4).

If they choose AID, semen from a donor, another man whom neither of the couple will ever know, is artificially put into the woman's vagina at the time of

328

ovulation. The unknown donor will be the father of her child. The child will inherit half its chromosomes from the unknown donor and half from its mother. It will inherit none from the man it regards as its father. The doctors will have made sure the donor neither suffers from cystic fibrosis nor is a carrier. But the child will be a carrier because it is certain to get one recessive cystic-fibrosis allele from its mother.

Adoption and AID are solutions likely to be available to the couple. The other solution, which is not generally available but may become so, is for the woman to have a test-tube baby: to be implanted with a donor egg cell fertilised by her partner's sperm (see Section 22.4). Then the unknown donor will be the mother of the child. The child will inherit half its chromosomes from the unknown donor and half from its father. It will inherit none from the woman it regards as its mother and who bore it and breast-fed it. The child will again be a carrier.

The couple may decide to have two children, one by AID and the other by implantation. One of their children will then inherit half the mother's chromosomes and the other will inherit half the father's chromosomes.

It is even possible to combine the two methods: the woman can be implanted with a donor egg cell fertilised with donor sperm. The child will then, like an adopted child, inherit none of the characteristics of the people it regards as its mother and father. Because the donors will have been carefully chosen, the child will not be a carrier.

There are two reasons why a couple might prefer double donoring to adoption. The first is that a woman would have the experience of being pregnant and of breast-feeding the baby, which would create a bond between them. The second is that doctors try to choose donors like the couple they are advising. Therefore a child born by double donoring would be more likely than an adopted child to be like a child born naturally to the couple.

Suppose a woman with cystic fibrosis wants to have a child by a man who neither has it nor has it in his family. They are unlikely to have a child with cystic fibrosis. They will do so only if the man is a carrier of the cystic-fibrosis allele. Even then the probability that they will have a child with cystic fibrosis is only 1 in 2 (1 to 1) because it is just as likely that the child will receive the father's dominant normal allele as his cystic-fibrosis allele. If the man is black, the genetic counsellor may say the chance of his being a carrier of cystic fibrosis is so slight that they can forget about it.

If the man is white, the genetic counsellor will recommend him to have a gene probe provided that he is in a country and a district where it is available. A gene probe will tell him whether or not he is a carrier. If he is not a carrier, there is no chance that the child will have cystic fibrosis. If he is a carrier, the probability that the child will have cystic fibrosis is 1 in 2 (1 to 1). The couple may or may not decide to have a child naturally.

If a gene probe is not available, all the genetic counsellor has to go on is the fact that between 1 in 20 and 1 in 25 white people are carriers of cystic fibrosis. The genetic counsellor will therefore tell the couple that the probability of their having a child with cystic fibrosis is between 1 in 40 (2×20) and 1 in 50 (2×25). The couple will almost certainly decide to have a child naturally.

If a couple decide to have a child knowing it may have cystic fibrosis, they may also decide to have the fetus tested (see Section 26.4) as soon as possible after the woman becomes pregnant. If the fetus has cystic fibrosis, it can be aborted. The mother can become pregnant again and allow a fetus to develop only if tests show it is normal.

The religious beliefs of some people forbid abortion. Other people do not like the idea of abortion. With the odds in their favour, such people will usually have children normally and take the risk that they may be born with cystic fibrosis.

Different ethnic groups are more likely to have some diseases than others. White people are more likely than black people to have both PKU and cystic fibrosis. Black people are more likely than white people to have sickle-cell anaemia. If a white person does not know of these diseases in his or her family, there is a probability of

1 in 80 that he or she is a carrier of PKU
1 in 20–25 that he or she is a carrier of cystic fibrosis

If a black person does not know of the disease in his or her family, there is a probability of

1 in 10 that he or she is a carrier of sickle-cell anaemia

If both parents are white and do not know of these diseases in their families, there is a probability of

1 in 25 600 ($80 \times 80 \times 4$) that their child will have PKU
between 1 in 1600 ($20 \times 20 \times 4$) and 1 in 2500 ($25 \times 25 \times 4$) that their child will have cystic fibrosis

If both parents are black and do not know of the disease in their families, there is a probability of

1 in 400 ($10 \times 10 \times 4$) that their child will have sickle-cell anaemia

All the calculations involve $\times 4$ because, even if both parents are carriers, the chance of having an affected child is only 1 in 4.

If one parent is white and the other black, they need scarcely consider the possibility that their child will have any of these diseases. Because PKU and cystic fibrosis alleles are rare in black people and sickle-cell anaemia alleles are rare in white people, it is most unlikely that a child of parents of different ethnic groups will have two of the same recessive alleles.

Fortunately more and more can be done to help people with inherited diseases or other handicaps by providing them with special training, specially adapted homes, special equipment and special motor cars. Such help is given both by the Government and by societies such as the Cystic Fibrosis Society. If, as a result of genetic counselling, fewer people inherit diseases and other handicaps, more money and resources will be available to give even better care to those who do

330 inherit them.

Q 26.1. Copy the table and put a tick ($\checkmark$) or ticks after each of the statements to show whether it is true of mitosis, of meiosis or of both. Either one or two ticks must be put on each line (a) to (g). The first line has been done for you.

	Statement	Mitosis	Meiosis
(a)	Forms only two nuclei from one nucleus	$\checkmark$	
(b)	Can form four nuclei from one nucleus		
(c)	Forms nuclei containing the same DNA		
(d)	Forms nuclei containing different DNA		
(e)	Takes place before human sperms can develop		
(f)	Can divide a diploid nucleus		
(g)	Can divide a haploid nucleus		

Q 26.2 A woman with cystic fibrosis has a child by a man who neither has cystic fibrosis nor is a carrier. Explain why it is certain that their child is a carrier of cystic fibrosis.

Q 26.3. Figure 26.5 shows a family pedigree. Clive was born with PKU. Diane is pregnant again.

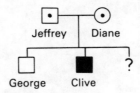

Figure 26.5   Family pedigree for Question 26.3

(a) Why might Jeffrey and Diane now go to a genetic counsellor?
(b) How will the counsellor know that Jeffrey and Diane are both carriers of PKU?
(c) What is the probability that the fetus has inherited PKU?
(d) (i) Why can the counsellor not know if George is a carrier unless he has a gene probe?
(ii) What is the probability that George is carrier?

Jeffrey   Diane

George   Clive

APPENDIX A: BACKGROUND IN MATHEMATICS AND OTHER SCIENCES

A.1 Mathematical background

To interpret the mathematical data that you collect in biological experiments you need only to add, subtract, multiply and divide. With these skills you can calculate percentages, arithmetic means, rates, ratios and probabilities.

(a) Fractions and percentages

A **fraction** is a number which is not a whole number. Examples of fractions are: $\frac{1}{2}$ or 0.5; $\frac{3}{4}$ or 0.75; $\frac{5}{4}$ or 1.25. These examples show the two different kinds of fractions: $\frac{1}{2}$, $\frac{3}{4}$ and $\frac{5}{4}$ are **vulgar fractions**; 0.5, 0.75 and 1.25 are **decimal fractions**.

When it is impossible to make a decimal fraction completely accurate (or when it takes a lot of figures to do so), it is usual to correct the number to two or three places of decimals. Thus $\frac{1}{3}$ may be written as 0.33 or 0.333 and $\frac{2}{3}$ as 0.67 or 0.667. Where the next figure will be 5 or more, as in 0.666, you raise the last figure by 1. You should be able to see that 0.67 is nearer to $\frac{2}{3}$ than 0.66.

A fraction is a proportion of 1. A percentage is a proportion of 100. To change a fraction to a percentage, multiply by 100.

$$\frac{1}{4} \times 100 = 25 \text{ per cent}$$
$$\frac{1}{2} \times 100 = 50 \text{ per cent}$$
$$\frac{48}{80} \times 100 = 60 \text{ per cent}$$
$$\frac{2}{1} \times 100 = 200 \text{ per cent}$$
$$\frac{11}{4} \times 100 = 275 \text{ per cent}$$

Suppose that baby A weighs 2.5 kg and baby B weighs 5.0 kg. Baby A's mass is half baby B's, while baby B's is twice baby A's. Baby A's mass is 50 per cent of baby B's, while baby B's is 200 per cent of baby A's.

If 23 of 31 seeds have germinated, what is the percentage germination?

$$\frac{23}{31} \times 100 = 74.2 \text{ per cent}$$

Both decimal fractions and percentages are useful when we want to compare different vulgar fractions. Suppose you want to compare the effect of different

333

temperatures on the rising of a flour-yeast dough over the same period of time (Experiment 5.3).

After 1 hour, the yeast-dough mixture at 25°C, whose height at the start was 37 mm, rose to 49 mm. Also after 1 hour, the yeast-dough mixture at 30°C, whose height at the start was 24 mm, rose to 32 mm. At which temperature was the rise proportionately greater?

Decimal fractions:

Rise at 25°C $= \frac{12}{37} = 0.324$

Rise at 30°C $= \frac{8}{24} = 0.333$

Therefore the rise at 30°C was greater

Percentages:

Rise at 25°C $= \frac{12}{37} \times 100 = 32.4$ per cent

Rise at 30°C $= \frac{8}{24} \times 100 = 33.3$ per cent

Therefore the rise at 30°C was greater

(b) Arithmetic means

To find the arithmetic mean or average of a set of numbers you add up the numbers and divide by the number of numbers. The mean of 11, 3, 9, 7 and 10 is

$$\frac{11+3+9+7+10}{5} = \frac{40}{5} = 8$$

The mean of 7, 12, 10, 9, 3 and 6 is

$$\frac{7+12+10+9+3+6}{6} = \frac{47}{6} = 7.83$$

To find the mean mass of strawberries produced by a particular growing method, you would weigh the strawberries produced by several plants and divide by the total number of strawberries. It is not necessary to weigh each strawberry separately: you can weigh all the strawberries together if you have a large enough weighing machine. The mean mass gives you a useful guide to the mass of each strawberry: it does not matter if none of the strawberries weighs that amount.

(c) Rates

When we say that someone's pulse rate is 72 beats per minute, we mean that it will beat 72 times in a minute. The number of times that something happens in a certain time is a rate. Suppose that in a photosynthesis experiment *Elodea* produces 850 mm^3 of oxygen in 78 minutes. What is the rate per minute at which oxygen is produced?

In 78 minutes the oxygen produced by *Elodea* is 850 mm^3

In 1 minute the oxygen produced by *Elodea* is $\frac{850}{78} = 10.9$ mm^3

Therefore *Elodea* produces oxygen at the rate of 10.9 mm^3 per minute.

If a pulse beats 33 times in half a minute (30 seconds), what is the rate per minute?

In 30 seconds the pulse beats 33 times

In 1 second the pulse beats $\frac{33}{30}$ times

In 60 seconds (1 minute) the pulse beats $\frac{33 \times 60}{30} = 66$ times

Therefore the pulse rate is 66 beats per minute.

Can you see how to do these sums? You begin by putting the information you have in a sentence that ends with the units (or whatever else) in which you want the answer.

Rates are often calculated in Biology: for example, volume of oxygen per minute, heartbeats per minute, breaths per minute, increase in number per day and increase in mass per year.

(d) Ratios

If a class contains 14 male and 16 female students, the ratio of males to females is 14 to 16 or 7 to 8 and we can write it as 7:8. The ratio does not tell us how many males and females there are. It tells us only the numerical relation between them: we know that for every 7 males there are 8 females. If there are 125 non-smokers in an office and 45 smokers, the ratio of non-smokers to smokers is 125:45 or 25:9 or 2.78:1. To express a ratio in the smallest possible whole numbers, divide the numbers by any common factors: the only factor common to 125 and 45 was 5. To express a ratio to 1, divide the larger number by the smaller. In the study of genetics, ratios approximating to 1:1 and 3:1 are common.

(e) Tables (tabulated data)

In a table, numbers can be listed both straight down, with a description at the top, and straight across, with a description on the left. A table can therefore give you two descriptions of every number, one above it and one to the left of it (see Tables 13.1 and 16.3). Remember that, if the numbers are measurements, the table must give the units of measurement such as grams (g) and metres (m). Whenever possible, give the units of measurement in the descriptions at the top or on the left. When this is not possible, put the units of measurement after the individual numbers in the table. All the units of measurement are given at the top in Table 16.1 and at the top and on the left in Table 16.3. In Table 16.4 all the units of measurement are put after the numbers.

When you design your own tables, you will usually be recording at least two sets of numbers. You will find it easier to put the sets of numbers in columns, that is to show them vertically, not horizontally. The first column should be the **independent variable**, the one you have been able to decide on for yourself. For example, in Experiment 2.1 you measure the diameter of the *Rhizopus* colony every day. You were able to decide on the day intervals and these are the independent variable. You cannot know what the diameter of the colony will be until you measure it. The diameter of the colony is therefore a **dependent variable**. The second and any further columns of your table are for the dependent variables.

Remember to head each column with a description of what you are listing and remember to put the units in the heading if you can.

Look at the table in Experiment 19.1: the first column (Time) is the independent variable; the second column (Pulse) is a dependent variable which you will be measuring in the experiment.

(f) Graphs

A **graph** is a diagram showing numerical values. Though it is not as accurate as a table, it makes it easier to grasp the relation between different values. It often makes it easier to see a trend. For example, if you graph temperature against time of day, you can see at a glance that temperature rises until the afternoon and then falls.

The term *graph* refers to the whole diagram; curves of best fit, and even straight lines joining points, are called **curves**.

Line graphs, bar charts, histograms and pie charts are different kinds of graph. In a GCSE Biology examination you might be asked to draw simple versions of any one of them.

(i) Line graphs

Most of the graphs in this book are line graphs. Two kinds of curves are used in line graphs; **smooth curves**, as in Figure 24.4, which shows height gains in boys and girls, and **jagged-line curves**, as in Figure 13.4, which shows nitrate levels in the River Lee. A smooth curve is a line of 'best fit' drawn *between* individual points and not from point to point on a graph. The jagged-line curve is a series of straight lines, drawn with a ruler, from point to point. You should always draw jagged-line curves yourself.

When you draw a graph, the horizontal axis (also called the *x* axis) is the line on which you space the **independent variable**, the one you have been able to decide on for yourself. For example, in Experiment 2.1 you measured the diameter of the *Rhizopus* colony every day. Time is therefore the independent variable and goes on the horizontal axis. Because you decided it for yourself, it is in regular whole numbers without fractions: 1 day, 2 days, 3 days, and so on. The horizontal axis should be labelled

Time in days

On the vertical axis (also called the *y* axis) you space the **dependent variable**, the one you have not been able to decide for yourself. In Experiment 2.1 it is the diameter of the colony. Because you have not decided it for yourself, it is in irregular numbers containing fractions: for example, 18.5 mm, 23.5 mm, 29.0 mm and so on. The vertical axis should be labelled

Diameter of colony in mm

In a GCSE Biology examination you are often told which set of numbers should go on which axis.

When you draw a graph, make sure that the scales you choose for your vertical and horizontal axes (plural of *axis*) cover (on the graph paper you are using) all the numbers you have to plot. You can use your graph paper either way round so that either the vertical or horizontal axis is on the longer side: choose the way round that allows you to use more of your graph paper.

After you have drawn your two axes, label each one with a description of what you will plot along it and of the unit in which it is measured unless it simply records numbers (such as the number of registered drug addicts on the vertical axis in Figure 21.2). If the numbers plotted on an axis are percentages, that must form part of your description (see Figures 22.4 and 24.5). A unit of measurement is best given as part of the description: for example, in Figure 24.4 the descriptions include the units: 'Age in years' and 'Height gain in cm per year'.

You should always draw both the axes and the curves in pencil but you may label the axes in ink. Write the numbers of the scales you have chosen against the darker lines on the graph paper. The point where the horizontal axis and the vertical axis meet need not equal 0 (see the vertical axes in Figures 13.1 and 19.10).

Suppose you have to draw the graph in Figure 11.8. Find the vertical line that is right for the independent variable (the number of weeks). Then find the horizontal line that is right for the dependent variable (the yield of the onion crop). Mark the point where the two lines meet. The best way to mark a point is to draw a dot in a circle ($\odot$): a dot without a circle is not easily seen. The second best way to mark a point is with a multiplication sign ($\times$): a plus sign ($+$) is likely to be hidden by the lines of the graph paper. Use $\odot$ as a general rule. Use $\times$ when you need a different point symbol for a different set of numbers: for example, see Figure 18.8.

When you have plotted all the points of your graph, join them in order by drawing straight lines between them with a ruler. This is what is meant by drawing a jagged-line curve. You must put your ruler on the two points but you should not draw your lines from point to point. Draw your lines to the edge of the circles and just short of the crosses. You can see how to do this by looking at the graphs in this book.

(ii) Bar charts

Bar charts are graphs in which one of the variables is not a number. Figure 16.6 is a bar chart in which one of the axes consists of the names of different countries while the other axis is numerical ('Deaths per 100 000 people'). A bar chart consists of narrow blocks of equal width. The blocks for the different countries in Figure 16.6 should not touch one another. If, however, Figure 16.6 showed birth rates as well as death rates, the two blocks for each country's birth and death rates should touch one another (but not any of the blocks for different countries). See Figure 18.10 for an example of a bar chart with touching blocks.

(iii) Histograms

Histograms are graphs in which numbers are shown in continuous groups. The groups are shown as blocks which must be touching. Figure 19.12 is a histogram in which the blocks have been organised according to the percentage of fat in the body (below 17.9, 17.9–21.6, 21.7–25.1, 25.2–29.7 and above 29.7).

(iv) Pie charts

Pie charts are graphs which show proportions of a whole. Because we are used to pies and cakes cut from the middle, we quickly grasp what proportion of the whole is in each segment. Pie charts start at 12 o'clock and usually go clockwise from the largest segment to the smallest.

Figure 13.3, showing the sources of radiation to which the average person in the UK is exposed, goes clockwise from the largest to the smallest segment. Figure 7.3, showing the sources of *Salmonella* infections, goes clockwise from largest to smallest except that the segment for unknown sources is put last.

In Figure 22.7 three pie charts show the different contraceptive methods used in rich countries, in poor countries other than China, and in China itself. The pie chart for rich countries goes clockwise from largest to smallest. So that 337 comparisons may be made easily, the other two pie charts follow the same order as the pie chart for rich countries.

(a) Solids, liquids and gases

Substances are **solid** when they keep the same shape (unless changed by outside forces). They are **liquid** when they have a definite volume and take up the shape of the bottom of their container. A **gas** moves freely through space, changing its volume to fill any container. Many substances, such as water, are a solid at low temperatures, a liquid at moderate temperatures and a gas at high temperatures. The change of a substance to gas, below the temperature at which it has to become a gas, is **evaporation**. For example, water becomes the gas steam at about 100°C; below that temperature water evaporates to become the gas **water vapour**.

(b) Energy

Energy is the capacity to do work. Energy is needed for activities as different as making cytoplasm and sending a rocket into space. In living organisms the important forms of energy are

chemical energy	the energy stored in chemical compounds
radiant energy	the energy from rays, such as rays of light
heat energy	the energy in heat
electrical energy	the energy of nerve impulses
kinetic energy	the energy of movement

(c) Temperature and humidity

The **temperature** of a substance is a measure of its heat energy. An increase in temperature increases the energy of the atoms or molecules of which the substance is composed: in a liquid or gas, therefore, the atoms or molecules move more quickly. Temperature is measured in degrees Celsius (°C).

Humidity is a measure of the water vapour present in air.

(d) Heat transfer

Heat energy is transferred by **conduction**, **convection** and **radiation**. When things are touching one another, heat is passed by **conduction** from places of higher to places of lower temperature. **Insulation** prevents heat from passing by conduction. Unit 20 describes how the skin loses heat by conduction from the warm blood to the cool surface. The fat layer below the skin insulates the warm body from the cool surface: it reduces but does not prevent loss of heat by conduction.

When heat has reached the skin from inside the body, it is lost to the air by conduction, convection and radiation. The loss of heat from the skin surface by the movement of air surrounding it is convection. (Any loss or gain of heat by the movement of a gas or liquid is convection.)

When the surroundings are very hot, it is possible for the skin surface to gain heat by conduction, convection and radiation, when the heat gained is passed to the blood by conduction. But all the sun and fires usually do for us is reduce the rate at which we lose the heat we have produced for ourselves by respiration.

(e) Solubility

Solubility is a measure of how easily one substance dissolves in another. A very soluble substance is one that dissolves easily. A liquid that has another substance dissolved in it is a **solution**. A liquid that has another substance dispersed in it, without being dissolved, is a **suspension**.

(f) Diffusion and diffusion gradients

Unit 7 describes the process of **diffusion**. It is the movement of ions or molecules from a region of their higher concentration (where there are lots of them) to a region of their lower concentration (where there are fewer of them). Diffusion goes on through liquids and gases until the ions or molecules are about equally spread out. Although they go on moving after that, they remain equally spread out. The movement from a more concentrated to a less concentrated region is along a **concentration gradient**.

Osmosis is a term used for the special case of the diffusion of water. It is the movement of water molecules from a region of their higher concentration to a region of their lower concentration. It usually involves movement through a membrane which allows water, but not dissolved substances, to pass through it.

A.3 Chemical background

An **element** is a substance that cannot be divided by chemical methods. A **compound** consists of two or more elements chemically combined. An **atom** is the smallest stable part of an element. A **molecule** is the smallest part of a substance that can exist by itself: the substance may be either an element or a compound.

An **ion** is an electrically charged atom or group of atoms. Simple inorganic substances often move into and through living organisms as ions. As ions, the parts of a compound such as sodium chloride (common salt) can move independently: for example, the chloride ions may enter a cell while the sodium ions stay outside it. **Salts** are compounds produced by the action of an acid and a chemical base (usually a metal): the compound sodium chloride is a salt.

pH is a measure of acidity and alkalinity related to hydrogen-ion concentration. The range is from pH 1 (very acid) to pH 14 (very alkaline); pH 7 is neutral.

Air consists of the gases nitrogen (about 78 per cent), oxygen (about 21 per cent), carbon dioxide (about 0.04 per cent), a variable amount of water vapour and some inert (inactive) gases. The oxygen in air is transported to living cells where it is used in aerobic respiration for the **oxidation** of compounds and the release of energy. **Combustion**, or burning, is the high-temperature oxidation of fuels involving changes of, for example, carbon to carbon dioxide, sulphur to sulphur dioxide and nitrogen to nitrogen oxides. It is one of the major causes of pollution of air.

APPENDIX B: INDIVIDUAL STUDIES

During the LEAG Biology (Series 17) course you do two Individual Studies that will be marked by your teacher as part of the GCSE examination. For each Individual Study you design and do an experiment and write a report about it. One of your experiments must deal with something in Part 1 of this book and the other with something in Part 2.

Each of your experiments should involve two to three hours' practical work. The two to three hours' practical work need not be done all at once. Having set up your experiment, you may need to leave it for a week before you record the results.

In each experiment you should measure at least one of the following: mass, length, volume of a liquid, temperature and time. Measurements should be made to the nearest 0.5 g, 0.1 cm (1 mm), 0.5 cm^3, 1.0 s and 0.5°C.

B.2 Choosing your experiments

You may not choose the Experiments in this book because you will have been taught how to do them. But you may choose experiments like the ones in this book. The following experiments suitable for Individual Studies are suggested in the sections on 'Interpretation of results' in this book:

> To test the hypothesis that temperature affects the growth rate of
> *Rhizopus* (Experiment 2.1)
> To test the hypothesis that carbon dioxide is taken up by a submerged
> *Elodea* plant in the form of hydrogencarbonate ions (Experiment 8.4)

You could test the hypothesis that any one of a large number of biological activities varies with temperature. You could test the hypothesis that a biological activity varies with pH, with the intensity of light or with the wavelength of light.

It is sensible to choose experiments that deal with parts of the Syllabus which you are interested in or which use practical equipment you like using and feel you have mastered. If you like cooking, you could choose an experiment about the rising of dough. You could do an experiment 'To test the hypothesis that the rising of dough is affected by adding vitamin C' or by some other ingredient or 'To test the hypothesis that yeast can use starch instead of sugar to make ethanol'.

You can be thinking of ideas for experiments all the time during your course and making notes about them. You may get ideas for experiments during theory classes. When you are interpreting the results of experiments you have done, you may get ideas for different but related experiments. You do not have to choose experiments done in a laboratory with test-tubes and beakers and Bunsen burners. You can do experiments with plants in your own garden or with dough in your own kitchen. You can do an experiment about pulse rates in your own home with no equipment except a watch that shows seconds.

You can do an experiment 'To test the hypothesis that pulse rate is affected by outside temperature'. If you were ill with a fever during your course, you could take the opportunity to do an experiment 'To test the hypothesis that pulse rate is affected by body temperature'. You can do an experiment 'To test the hypothesis that pulse rate decreases with age'. For that matter you can do an experiment 'To test the hypothesis that pulse rate increases with age'. You might settle for an experiment 'To test the hypothesis that pulse rate is affected by age'.

You could get the same marks for any one of these three experiments. It does not matter whether or not the hypothesis you test turns out to be true. Science progresses by gaining knowledge. Knowledge that something is untrue is often as important as knowledge that something is true.

When you have an idea for an experiment, ask your teacher if it is suitable for an Individual Study. Do not be disheartened if you have to suggest several ideas before you are told you have a suitable one. When you do have a suitable idea, you can set about planning your experiment and gaining marks. Your teacher will mark each of your Individual Studies out of 25 using the scheme below.

Planning the experiment	5 marks
Reporting the methods	5 marks
Presenting the results	5 marks
Interpreting the results	5 marks
Evaluating the experiment	5 marks
	Total 25 marks

B.3 Planning an experiment

Before you do any practical work, you must submit a written plan of your experiment to your teacher. Your teacher needs to see this to make sure that your experiment is likely to produce suitable results within the two to three hours' practical work, that it can be carried out safely and that it is not too difficult for you.

Your teacher also marks your plan out of five. It may be that your plan is unsuitable. If so, your teacher will tell you why and will give you the chance to make changes. Only if your teacher has to help you make the changes will you lose some of the available five marks. Do not be disheartened if you lose a mark or two at this stage. It is worth it to start with a good plan and, as you know, there are 25 marks available for each experiment.

First your plan must state clearly what the experiment is. It should begin: 'To test the hypothesis that...' It should describe how you will do the practical work. You may use a flow chart to do this or you may list the steps in your method. Whichever way you do it, you must give the same kind of detail as is given in Experiments in this book (including detail that is necessary for the sake of safety). Suppose you are doing an experiment to test the hypothesis that carbon dioxide is taken up by a submerged Elodea plant. It is not enough to write:

WRONG

Put a bit of pondweed in a beaker of water and shine a light on it. Count the number of bubbles coming out of the plant and record it in a table.

You need to write:

RIGHT

Cut 5 cm off the end of a piece of pondweed which has bubbles coming out of it. Put it into a beaker which has 100 cm^3 of cold tap water in it.
Measure 50 cm away from the beaker and put a lamp with a 40-watt bulb there. Plug in and switch on the lamp with dry hands.
Every five minutes for half an hour count the number of bubbles coming out of the plant in 30 seconds and record it in a table.

The more measurements the better. Take as many as you have time for.

Also, like the Experiments in this book, your plan must include a list of any equipment, chemicals and biological material that you will use.

(a) Repetition

Your plan should say you will repeat your experiment at least five times unless that is impossible within three hours. If necessary you can do five identical experiments at the same time. Can you see why you should repeat your experiments?

Tossing a coin in the air once would not tell you whether or not it was weighted. Similarly, doing an experiment once is not a reliable way of testing a hypothesis. Repetition is necessary to recognise and rule out fluky results. When an experiment is done in class, it is repeated by lots of different students. When you do an experiment on your own, you must do the repetition yourself. If you are testing the hypothesis that the rising of dough is affected by adding vitamin C, you should use several identical samples of dough. If you are testing the hypothesis that people with large hands tend to have large feet, you should measure at least five people's hands and feet.

(b) A control experiment

Suppose you are testing the hypothesis that temperature affects the growth rate of *Rhizopus*. To recognise such an effect, you must know what happens to the growth rate of *Rhizopus* if the temperature is not changed. Suppose you measure the growth rate of five samples of *Rhizopus* at 10°C below room temperature. 343

You must also measure the growth rate of five samples of *Rhizopus* at room temperature. This is the control experiment done for comparison. A control experiment is one in which you keep unchanged whatever you change in your main experiment.

B.4 Records

From now on keep everything you write down about your experiment. Keep your plan when your teacher has approved it. If your plan has been changed, keep all the different versions of it. Make and keep records of everything you do in your experiment. Note when the experiment goes differently from the plan. Make records of all the results you obtain. Your results must be submitted, unaltered, with your report even if you spill tea on them (which is a reason for not spilling tea on them).

Although no one need see you doing your experiment, you must be able to provide evidence that you did it yourself. You will be interviewed about your experiment by your teacher and perhaps by an Assessor from LEAG. Provided you make and can show a record of all you did in your experiment, you will be in no difficulty.

B.5 The experiment

Once your teacher has approved your plan, you may do your experiment as soon as it is convenient. Before you begin your experiment, you must decide how you will record your results and you must draw up any tables you will record them in. Have a watch or clock with you.

Once you have begun your experiment, record briefly everything of importance and the time when it happens. It is best to note all the times on the left-hand side of the page. Your brief notes will be the basis of the report you write later. Record both what you do yourself and what happens as a result. Record both what you expected and what you did not expect. Do not think your experiment has to go as you expected for you to get good marks. What matters is that you show you are an accurate observer and reporter and that you say something sensible about anything unexpected that happens. Above all, take your measurements carefully and record them carefully.

B.6 The report

Although you must hand in both your report and the original record of your results, it is your report on which your experiment is judged. It should be headed 'To test the hypothesis that...' Next you should say why the information you hoped to get from the experiment would be worth having. Suppose your experiment is to test the hypothesis that temperature affects the growth rate of *Rhizopus*. You know *Rhizopus* is a fungus that makes food go bad (see Figure 344 7.1). If your experiment shows that *Rhizopus* grows less quickly at a lower temperature, that will be a reason for keeping foods affected by *Rhizopus* at low temperatures. It will also suggest that compost will form more quickly in hot weather.

Next write a straightforward account of how you did your experiment. If you have done the *Rhizopus* experiment, your report could begin:

> I took a closed sterile Petri dish. I turned it over and drew two lines at right angles across the middle of its base with a fine marker pen. I wrote the time and date near the edge of the bottom of the dish.

If you have done the experiment to test the hypothesis that carbon dioxide is taken up by a submerged *Elodea* plant, your report could begin:

> I cut 5 cm^3 off the end of a piece of pondweed which had bubbles coming out of it. I put it in a beaker which had 100 cm^3 of cold tap water in it.

This part of your report will closely follow your plan. But do not be afraid to say if your experiment went differently from your plan, whether it was because you thought of a way of improving it or because something went wrong. If you improved it, explain how.

Even though you also hand in the original record of your results, your report must present your results in a clear and tidy form. It must show too any calculations you have made from your results. Draw a graph or diagram (or both) if it makes the results themselves or any conclusion you draw from them easier to understand.

Needless to say, the conclusion (or conclusions) you draw are important. If you have done the *Rhizopus* experiment and it has worked as you expected, you will conclude that *Rhizopus* grows less quickly at 10°C lower than room temperature. You can say that this suggests that *Rhizopus* may always grow more slowly at a lower than a higher temperature. Do not overstate your case. Do not say you have proved that all foods keep better at lower temperatures. Say only what you are entitled to say on the evidence of your own experiment. Avoid the word 'proved' altogether. We say something is proved only when an experiment has been repeated successfully many times.

Last but by no means least, your report must evaluate your experiment. That does not mean saying only what it has achieved. If there were things wrong with it, you will get marks for pointing them out. You can begin by pointing out that you were not able to repeat the experiment as often as you wanted. If your experiment went differently from your plan without your having made any deliberate changes, explain why as best you can. It may be because you did something wrong. It may be that you did everything as you had planned yet the unexpected happened. If so, try to suggest why.

Say, if you can, how you could improve the experiment with the knowledge you have gained from doing it. Say, if you can, how you could improve it if you could use more and better equipment (such as more accurate measuring equipment). If possible, suggest experiments which would give more information about the hypothesis you have tested. If you have done the *Rhizopus* experiment, you can say you would test the hypothesis further by measuring the growth rate of *Rhizopus* at many different temperatures.

APPENDIX C: HOW TO DRAW; HOW TO COMPARE BIOLOGICAL SPECIMENS; HOW TO ANSWER EXAMINATION QUESTIONS; HOW TO REVISE

C.1 Assessment of practical work

In the LEAG (Series 17) examination 20 per cent of the total marks are given for practical work. Your own teacher will mark your practical work out of 80 using the scheme below.

Section 1		Total
Drawings of biological material	10 marks	
Comparisons of biological material	10 marks	
Measurements with apparatus	5 marks	
Procedure and use of apparatus	5 marks	30 marks
Section 2		
Two Individual Studies	25 marks each	50 marks
		80 marks

Appendix B helps you with your two Individual Studies. The first part of this   347

Appendix tells you how to draw and how to compare biological specimens. As you can see from the scheme above, you will also be given marks for taking measurements accurately, for going about your experiments methodically and for using apparatus correctly (and safely): these are abilities that you will develop with practice under your teacher's guidance.

You do not need to be an artist to do good biological drawings. But you must be accurate and your lines must be clear and tidy. Biological drawings are not pictures: do not shade them to give them a solid look. Draw the outlines of structures to show their positions and shapes accurately. Use a hand lens to look at the details on hand-held specimens. In Experiment 8.1, for example, it is better to draw accurately only one or two cells of each type (and to leave the rest undrawn) than to draw a lot of cells inaccurately. Always draw in pencil because you will have to rub out many times if you are going to be accurate. It is easier to be accurate in a large drawing and it does not matter if it leaves a lot of unfilled space. Each finished drawing should take up about half a page.

Writing in the names of parts is called **labelling** a drawing, and the names of the parts are **labels**. A line from the label to the part is called a **labelling line**: it should be drawn neatly in pencil with a ruler. Two labelling lines should never cross each other. Print your labels horizontally so that they are easy to read. Print them in pencil so that, if they are wrong, you can rub them out.

Whenever you draw biological material you should try to give the scale of your drawing. A drawing of a leaf and of a *Rhizopus* colony both take up half a page. Without scales to help them, other people may have no idea of the size of what you have drawn.

The simplest scale you can give is a straightforward measurement. If it is a hand-held specimen, use a ruler to measure its actual length, width or diameter. Record your measurement at the side of your drawing. Here are two examples of how to do this:

length of leaf = 73 mm
diameter of colony = 28.5 mm

If the specimen has an irregular shape, it is best to draw a two-headed arrow (↔) across the part you have measured and record the length of the arrow at the side of your drawing:

arrow length = 19.5 mm

There are other ways of showing the scale of your drawing. One is to show its magnification in brackets (× 10) means the drawing is ten times larger than the specimen; (× 0.1) means the drawing is ten times smaller than the specimen. Many of the Figures in this book give scales in this way. But you will find it easier to give the measurement of a specimen across one part.

How do you show the scale of a microscope drawing? If the magnification of

348

the eyepiece is ×10 and the magnification of the objective lens is ×4, the magnification of what you see is ×40. But this is not your scale because what you draw is not the same size as what you see. What you draw needs to be bigger in order to be clear.

If you are looking at a specimen on a microscope slide, use a micrometer graticule to measure it as you would use a ruler (see Experiment 8.1). A micrometer graticule is like your ruler except that it is only 10 mm (1 cm) long and is numbered in 1 mm intervals subdivided into 0.1 mm intervals. Write the scale at the side of your drawing in the usual way:

width of leaf = 2.6 mm

Under high power it is too difficult to use the graticule to measure what you see. Do not therefore try to give the scale when you draw under high power, which in practice means when you draw cells.

Remember these rules:
○ use an HB or a B pencil with a sharp point
○ be accurate, not artistic
○ take up about half a page with each drawing
○ use a ruler to draw labelling lines
○ do not cross labelling lines
○ print labels horizontally
○ give a scale

C.3 How to compare biological specimens

When you compare two biological specimens, you must try to find at least five important ways in which they are similar or different. The best way to show the similarities and differences is to make a table by dividing your page vertically into two halves. Head each half with the name of one of the specimens. Suppose you are given a parsley plant and a mint plant to compare (see Figure 9.1). A table would look like this:

Parsley	Mint
1 Leaves deeply divided	Leaves undivided with a jagged edge
2 Leaves grow from base of plant	Leaves grow from stem in pairs
3 Long leaf stalks	Short leaf stalks
4 One long thick root with thin roots on it	Many short thin roots
5 No stem seen	Stem thick and ridged

C.4 How to answer examination questions in the written papers

The most important thing is to read the questions carefully. If you do not answer the question you are asked, what you write may be correct yet get you not a single mark. Suppose you are asked what happens to the muscle at the back of   349 your thigh when you bend your knee. You may feel so pleased because you know the name of the muscle that you write down 'Biceps'. If you do, you will get no mark. You have not been asked to name the muscle. You have been asked what happens to the muscle when you bend your knee. The answer is 'It contracts'.

On the other hand, you do not lose marks for giving wrong answers. If you do not know an answer, you should try to work it out or even guess it. Suppose you do not know what happens to the muscle at the back of your thigh when you bend your knee. You probably know that muscles both contract and relax. The answer is probably either 'It contracts' or 'It relaxes'. Try to work out which is the more likely. Whatever you do, write down one or the other but not both.

(a) Diagrams

Always draw a diagram when the question asks you to. Even if the question does not ask for a diagram, you may be able to answer it more clearly and more quickly with a diagram on which you write notes as well as labels. If you give information in notes on a diagram, you need not repeat it in your written answer. Because time is so important in an examination, do not waste it by drawing a diagram that does not help to answer a question.

(b) Allocating your time

You must not spend so long on some questions that you do not have time to answer all the questions you are supposed to answer. If you find a question difficult, that means you are not sure of the answer to it and may not be able to get any marks for it. It does not make sense to spend a long time on a question like that. You can spend the time better answering questions which you find easy or less difficult and on which you can expect to get marks. Suppose the last question has been given 12 marks and you do not leave yourself enough time to answer it. That is 12 possible marks you have lost. You have almost certainly lost some easy marks because nearly every long question begins with some easy bits.

If an examination paper is 2 hours (120 minutes) long and there are 100 marks given to the questions in it, you can work out that you have just over a minute in which to get each mark. If an examination paper is $1\frac{1}{2}$ hours (90 minutes) long and there are 60 marks given to the questions in it, you can work out that you have a minute and a half in which to get each mark. Then, when you see the marks that are given for a question, you have a rough guide to how long you should spend on it. Suppose you have just over a minute in which to get each mark. If a question has 3 marks given for it, spend up to three minutes answering it; if a question has 8 marks given for it, spend up to eight minutes answering it.

Because you have just over a minute in which to get each mark, that will leave you some time over at the end. Because you can expect to answer some questions in less than the time you allow for them, you will probably have even more time over at the end. Use any time you have over to go back to the questions you left unfinished. If you finish all the questions before the time is up, read through your answers to see if you can improve any of them.

350   Even if the room where you will sit the examination has a clock, take your

own watch so that you can see the time without looking up. Besides, the clock in the room may go wrong.

(c) How much should you write?

Do not write more than you need to in order to answer the question or you will waste time. In some papers you are given spaces in which to write your answers. The space you are given for each question is the most that can be needed by someone with large handwriting.

Suppose that the question is:

What word describes the microorganisms that feed on dead organic material?
(1 mark)

It is enough to write the one word 'Saprophytes'. It is a waste of time to write: 'The word that describes the microorganisms that feed on dead organic material is saprophytes.'

Just as the marks for each question give you an idea of how long to spend on it, so they give you an idea of how much to write. Where there is only 1 mark for a question, it is usually enough to write one word or a few words. Where there are 4 marks for a question, the examiners expect at least four different points. Suppose that the question is:

Why do you think nitrate levels in rivers in south-east England have risen during the last forty years?
(4 marks)

Suppose you write: 'Because of fertilisers.' That answer is likely to get you only 1 mark, for implying correctly that nitrates come from fertilisers. You need to make some more points: that fertilisers have been used increasingly during the last 40 years (1 mark) because fields have been kept growing crops continuously (1 mark) with the result that inorganic ions have not been replaced naturally (1 mark); that fertilisers have also been used increasingly because new varieties need more (1 mark); that fertilisers dissolve in rain and drain into rivers (1 mark). Make at least four points if you can. It is a good idea to make five points in case the examiners do not accept one of yours. But do not go on to make six, seven or eight points. It all takes time you can spend better on the other questions. If you have time over at the end and have finished all the other questions, you can come back to this question and make some extra points.

(d) Scientific terms

Use scientific terms if you know them but explain things any way you can if you do not. For example, it is better to use the term *vasodilation* if you know it, but better to say *widening of the arteries* than to say nothing.

(e) Things you have not learnt about

Do not expect all the questions to be about things you have studied. The GCSE examination tests what you understand as well as what you know. Some of the questions tell you things you are not expected to know and ask you to explain them. You explain them by applying what you know to the information given you in the question and by using your common sense. You must read such questions carefully to make sure you understand them and will be able to use

351

information given in them. You are unlikely to be given much information you do not need to use. Do not be surprised if you need to read such questions two or three times: they will make more sense each time you read them.

(f) Designing experiments

By the end of the course you will have had practice in designing experiments to test hypotheses and in writing reports of what you did. Give numbers and measurements whenever you can. Do not say 'a large number of seeds'; say '50 seeds'. Do not say 'after some time'; say 'after 3 hours'.

(g) Past examination papers

Many of the questions at the end of the Units are like those in the examination. To get more practice you should answer questions from previous years' examination papers.

(h) Examiner's reports

Many GCSE Groups publish reports on each year's examinations. You can learn a lot from these reports and the past papers to which they refer. The reports explain what answers were expected and what the common mistakes were.

C.5 How to revise

When you revise for the examination, remember that you learn best by making yourself think. Do not just read through this book or your notes. Look at the titles of the Units and of the main Sections. Ask yourself what you know about each topic; make a few rough notes and drawings to jog your memory. Only after you have done all that should you go through the Unit to check that what you have remembered is correct and to see what you have missed out. In this way you discover the topics that you really need to revise. Make a list of them all and revise them. When you have done so, work your way through the list of topics you needed to revise, seeing what you can remember of each one and only looking back at the Units after you have done so.

Revision is not something that need wait until the examination is near. You can revise the Units you have studied at any time. The better you understand and remember the Units you have already studied, the easier and more interesting will be those you study in future.

352

APPENDIX D FOR TEACHERS: PRACTICAL WORK

Laboratory experience shows that all the 23 Experiments in this book can be done by students. Some students will need help in following instructions, in manipulating apparatus, in reading, recording and interpreting results, and, not least, in designing experiments. Even when you are assessing students, you may give them help if you take marks off for it; indeed you may modify an Experiment to make it easier if you take marks off for doing so (see the LEAG Syllabus under Appendix B, Teacher Assessment of Practical Work).

You are in fact expected to help students and make experiments easier for them. The joint DES and Welsh Office document *GCSE The National Criteria: Biology* (HSMO, 1985) declares (our italics): '*It is essential that assessment schemes are devised to enable candidates to show what they can do rather than what they cannot do.*' It may hurt to take marks from your students before they have begun but, if they are going to have difficulty, it is the way to maximise their marks. Students trying to do practical work they do not understand are likely to get few marks. Bear in mind that research has shown we all tend to overestimate our own students' practical skills. Before modifying experiments to make them easier for students, you should read the booklet *Science GCSE: A Guide for Teachers* (Open University Press in conjunction with the Secondary Examinations Council, 1986), which examines the problems of setting differentiated tasks. You should read also *Underway with GCSE Biology* (*Mature*) (LEAG, 1987), which gives teachers advice on assessment.

Table D.1 shows which practical skills required by LEAG can be assessed by each of the Experiments in this book.

Experiments 4.2, 5.2, 6.3 and 18.1 require students to design experiments themselves, though with some guidance in the text. Just how they do this is a matter for your judgement and must depend on the standard the students have reached at the time of each experiment. The class can design an experiment collectively under your guidance. Individual students or pairs or groups of students can each design an experiment during a practical class with the option of turning to you for help. Or you can set designing an experiment for homework. Designing these experiments is practice for the students' Individual Studies.

Except in these four 'design' Experiments and in Experiments where the 'Instructions to students' indicate otherwise, students should work individually, even if it means sharing equipment, so that their skills can be assessed.

353

Table D.1 The practical skills assessable during each Experiment

Experiment	1.1 Making drawings	1.2 Making comparisons	1.3 Measurements with apparatus					1.4 Procedure
			Mass	Length	Volume	Time	Temperature	
1.1		✓						✓
2.1			✓					✓
3.1					✓	✓		✓
4.1			✓			✓		✓
4.2								✓
5.1					✓			✓
5.2								✓
5.3			✓		✓		✓	✓
6.1			✓		✓			✓
6.2							✓	✓
6.3								✓
7.1	✓							✓
8.1	✓							✓
8.2								✓
8.3			✓					✓
8.4			✓	✓				✓
9.1	✓				✓			✓
14.1	✓							✓
15.1								✓
15.2					✓	✓		✓
16.1			✓		✓			✓
18.1								✓
19.1						✓		✓

The Experiments in this book assume: a laboratory with mains electricity, gas, mains water, hand-washing facilities and sufficient work space for a class; access to a refrigerator, to an incubator and to storage space where students can leave experimental material for up to two weeks; an autoclave or high-dome pressure cooker to sterilise glassware and reagents for class use and to dispose aseptically of contaminated material; and the equipment, chemicals and biological material listed in the Experiments. Distilled water is required in quantities that justify a still.

The lists constitute what is desirable rather than what is essential: jars can be substituted for beakers, watches for stopclocks; one microscope, balance or Bunsen burner can be shared by a number of students (at the cost of making experiments last longer). However much you have to economise on equipment, you should provide students with micrometer graticules for the direct measurement of material under microscopes (LEAG practical skill 1.1). These graticules (which can be placed on coverslips) are 10 mm long, numbered at 1 mm intervals, with 0.1 mm subdivisions. Without micrometer graticules, students find it almost impossible to give correct scales for their microscope drawings.

Nor can there be any economising in safety equipment. Suitable dilutions of disinfectants must be available whenever students are handling microorganisms. Spillages must be cleared away aseptically and immediately. Safety spectacles must be worn whenever liquids are heated; both safety spectacles and gloves must be worn whenever hazardous chemicals are handled. Solutions must be pipetted with a pipette filler or teat, not by mouth. Passages affecting safety are in tinted red boxes.

You should read *Microbiology*: *An HMI guide for schools and non-advanced further education* (HMSO, 1985) before allowing students to work with *Rhizopus* (Experiments 2.1 and 7.1).

Students can perform some of the simpler experiments at home.

INDEX

Bold type indicates relative importance. The index does not refer to Questions, Practical Work or Appendices.

web 142
freezing 86–7
fruit 115, **117–18**
fungicide 86, **148**
fungus 7, 15–20, 82–3, 85–6, 88–90

gamete 115, 117, 167, 263, 321, *see also* egg
      cell, sperm
gas exchange 216–17
gene 309–17
   probe 326, 329
   reserve 154
genetic
   counselling 317, **328–30**
   engineering 8, 83, 89, 148, 309, **315–17**
   fingerprinting **308**, 324
   variation 129, 323–4, 326–8
genetics 305–19
genital herpes 275
genotype **312–14**, 327
genus 5
German measles *see* rubella
germination 130
gestation *see* pregnancy
gland
   adrenal 248, 254
   ductless 254
glasshouse 131–3
glucose 53–7, 95–6, 100, 173, 207–8
glue-sniffing 258
glycogen 15, 173, 207–8
gonorrhoea 275
grape 68–70
grass 135
growth 16–19, **31–4**, 56, **297–9**
   curve 32–3, 56–7
gut 187–91

habitat 153–4
haemocytometer 34–5
haemoglobin 166, 216
haploid 321–2
hay fever 48, 116
heart 227, **229–37**
   attack 233
   disease 193–5, 233, 236–7
heat
   loss and gain 241–5
   transport 242
height 327
hepatitis 257
herbicide 148
herbivore 141
heroin 256–7
heterotropic 81
heterozygous 312
HIV 275
homeostasis 246
homozygous 312
hops 66–7

hormone 254–5, 266, 271, 274, 279–80, 299–300
hydrogen peroxide 41–3
hydroponics 132–5
hypha 17–20
hypothermia 245–6
hypothesis 42

implantation 279–80
in vitro 46–9, 131
   fertilisation 266–7
incubate 21
induction 294–5
infertility 266–7
inheritance 321–30, *see also* genetics
inoculate 20
inoculum 21, 25
insecticide 146–8
insulation 243–4
insulin **89**, 316
intervertebral disc 209–11
intra-uterine device *see* IUD
ion **100–1**, 123–5, 132, 135, 159–60, 173,
      177–8, 249–50
iris 253–4
iron 177
irradiation **23–4**, 86–7
IUD 272

karyotype 321
key 8–12
kidney 247–50
knee 203–5
kwashiorkor *see* protein deficiency disease

labour 293–5
lactation 191–2, *see also* breast-feeding
lactic acid 54–5, 65, **71–3**, 206–7
Landsaver 134–5
leaching 124
leaf **97–100**, **113–115**
life
   cycle 115–18
   expectancy 302
lifting 211–12
ligament 205, 209–10
light 95–8
limiting factor 32–3
lipid 15, 147, **173–5**, 188–9, 233
liver 173, 190, 207–8, 247, 258
lung 215–23

magnesium 101
magnification 10
malt 65–6
mammal 3
mammary gland 3, 296
medicine 256
medium 20–5
meiosis 321–4
menopause 300

menstrual cycle **265–6**, 299
menstruation 265–6
mesophyll 98
metabolism 41, 57
microorganism **15–20**, 146, *see also*
    compost, food spoilage, nutrition
micropropagation *see* tissue culture
milk 296
mitosis **305–7**, 322
mould *see* fungus
movement 203–8
mucus 220–1, 223, 317
multiple alleles 313
muscle
    antagonistic **203**, 253
    contraction **206**, 253
    energy 206–8
    gut 187
    skeletal 203–8
    structure 205–6
mutagen 327
mutation 325–7
mycelium 17–20
mycoprotein 7, **90–1**, 176

natural selection 137, 148
nectar 116, 145
negative feedback 246
nerve impulse 206
nicotine 223
nitrification 123–5
nitrogen 100–1
    cycle 123–5
    fixation 124–5
nucleus 15, 17, 97, 165, **305–8**, 321–3
nutrition
    microorganism 81–6
    plant 95–101

obesity 195
oestrogen 271, 299–300
oil 156–7, *see also* lipid
optimum 44–5
organ 111, 168–9
organic 81, 83, 101
organism 3, 111
ovary 263–6
oviduct 264–7
ovulation 264–5
ovule 115–18
ovum *see* egg cell
oxygen *see* photosynthesis, respiration
    debt 207, 219

pain killer 256
parasite 84–6
pathogen 85
pedigree 317–19
penicillin 7, **88–9**, **256**, 275
Penicillium 7, 20, 83, **88–9**
peristalsis 187

pest **142–4**, 146–7
pesticide 142–4, 146–9, 159
pH 32, 45
phenotype **312–14**, 327
phenylketonuria *see* PKU
phloem 113
photosynthesis **95–100**, 122–3
phototropism 96–7
pickling 86–7
PKU **295**, **312**, 330
placenta 280–2, 290
plant 5, **111–18**
    breeding 137–8
plasma *see* blood
plasmid 17, 309
platelet *see* blood
pollen 115–17
pollination **116**, 145
pollution 149, **155–60**, 219–22
population size 267–9
post-natal screening 295
precipitation 121–2
pre-embryo 280
pregnancy 191–2, **279–90**
producer 95, 141, 149
prokaryote 7, 17
propagation 130–1
protein 42, 100, 123, 173, **175–7**, 188–9
    deficiency disease 176–7
    digestion 47, 188
protoctist 6
puberty 299–300
pulse 231–2
pupil reflex 253–4
putrefaction *see* decay
Pythium 85–6

radiation 87, **157–8**, 327
recessive allele **311–14**, 317
recombinant DNA 309
recycling **121–7**, 146
    carbon 121–2
    nitrogen 123–5
    paper 154–5
    water 121–2
reflex action 253–4, **295**
refrigeration 86–7
rennin 46–7
reproduction 16–19
    asexual 118, 130
    sexual 115–18, 129
respiration 43, **53–7**, 122–3, 191, 206–7
    aerobic **53–7**, 206–7
    anaerobic **53–7**, 206–7
response 253, 258–9, 295
    shoot 96–7
Rhesus factor 289
Rhizopus **17–19**, 33, 82–4
rhythm method 273
rickets 178

360

root   111–13
roughage   173, **179–80**
rubella   284–5

Salmonella   84–5
saprophyte   **81–5**, 122–3, 125–7
scrotal sac   263
scurvy   178
secondary sexual characteristic   299
seed   115, **117–18**, 129–30
selective breeding   5, 118, 138
selective reabsorption   249
semen   168, 263–4
senile dementia   *see* Alzheimer's disease
septicaemia   257
sex
   chromosome   324–5
   inheritance   324–5
sexually transmitted disease   *see* STD
sickle-cell anaemia   **318–19**, 330
silage   54, **73–4**
single-cell protein   8, 90–1, 176
sitting   210–11
skin   241–7
smoking   **222–4**, 234, 236–7, **285–7**
solvent   258
spacing   135–7
species   5
sperm   **167–9**, 263–7, 271–3, 299, 323–5
spirometer   217–19
spore   **18–24**, 85–6
stamen   115
standing   210
starch   100, 173
STD   272, **274–6**
stem   113
sterilisation   **21–26**, **272–3**
stoma   99–100
substrate   43, **53–7**
sulphur dioxide   155–6, 220–1
support   209–11
sweat   243
syphilis   275

target organ   254
technology   48–9
temperature   32, 44–5
   control   241–6
tendon   205
testis   263
testosterone   299–300
test-tube baby   266–7
thalidomide   287–8

thrombosis   *see* blood clot
tissue   111, 168–9
   culture   131–2
tongue-rolling   311–12
toxic substance   250
transpiration   **112–13**, 121–2
triglyceride   **174–5**, 207–8
turbidity   37
twin   263

ultrasound scan   282–3
umbilical cord   280–2
urbanisation   153
urea   190, **247–50**
urine   **248–50**, 289
uterus   264–5, 279–84

variation   129, 323–4, **326–8**
variety   118
vascular
   bundle   96
   tissue   113
vasectomy   272
vasoconstriction   244
vasodilation   243–4
vector   143
vegetative propagation   *see* cloning
vein   96–8, 113, **227**, **229–32**
ventilation   *see* breathing
ventricle   230–2, 234–5
vertebral column   4, **209–11**
villus   189, 280–3, 290
vinegar   54, **70**, 87
virus   275, 284–5
vitamin   173, **178**

waste removal   246–50
water   31, 68, **96–8**, **111–13**, **121–2**, 173,
   178–9, 247–50
weed   136–7
   killer   *see* herbicide
weight   288–9, *see also* obesity
wine   68–70
withdrawal symptom   259

X-ray   157, 327
xylem   98, 113

yeast   7, **15–17**, 32–7, **54–7**, **67–9**, 82–3,
   90–1
yoghurt   54, **70–2**

zygote   115, 117, 264, **280**